John F. Blake

Astronomical Myths

John F. Blake

Astronomical Myths

ISBN/EAN: 9783337183073

Printed in Europe, USA, Canada, Australia, Japan

Cover: Foto ©berggeist007 / pixelio.de

More available books at **www.hansebooks.com**

ASTRONOMICAL MYTHS,

BASED ON

FLAMMARION'S

"HISTORY OF THE HEAVENS."

BY

JOHN F. BLAKE.

London:

MACMILLAN AND CO.

1877.

LONDON :

R. CLAY, SONS, AND TAYLOR, PRINTERS,

BREAD STREET HILL,

QUEEN VICTORIA STREET.

PREFACE.

THE Book which is here presented to the public is founded upon a French work by M. Flammarion which has enjoyed considerable popularity. It contained a number of interesting accounts of the various ideas, sometimes mythical, sometimes intended to be serious, that had been entertained concerning the heavenly bodies and our own earth; with a popular history of the earliest commencement of astronomy among several ancient peoples. It was originally written in the form of conversations between the members of an imaginary party at the seaside. It was thought that this style would hardly be so much appreciated by English as by French readers, and therefore in presenting the materials of the French author in an English dress the conversational form has been abandoned. Several facts of extreme interest in relation to the early astronomical myths and the development of the science among the ancients having been brought to light, especially by the researches of Mr. Haliburton, a considerable amount of new matter, including the whole chapter on the Pleiades, has been introduced, which makes the present issue not exactly a translation, but rather a book founded on the French author's work. It is hoped that it may be found of interest to those who care to know about the early days of the oldest of our sciences, which is now attracting general attention again by the magnitude of its recent advances. Astronomy also, in early days, as will be seen by a perusal of this book, was so mixed up with all the affairs of life, and contributed so much even to religion, that a history of its beginnings is found to reveal the origin of several of our ideas and habits, now apparently quite unconnected with the science. There is matter of interest here, therefore, for those who wish to know only the history of the general ideas of mankind.

THE ANNUAL REVOLUTION OF THE EARTH ROUND THE SUN, WITH THE
SIGNS OF THE ZODIAC AND THE CONSTELLATIONS.

LIST OF ILLUSTRATIONS.

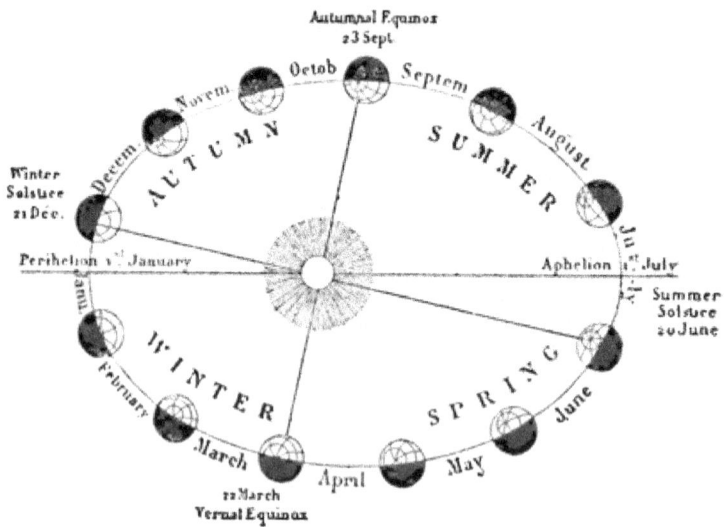

THE EARTH'S YEAR, AND THE MONTHS.

CONTENTS.

HISTORY OF THE HEAVENS.

CHAPTER I.

THE FIRST BEGINNINGS OF ASTRONOMY.

ASTRONOMY is an ancient science; and though of late it has
made a fresh start in new regions, and we are opening on
the era of fresh and unlooked-for discoveries which will soon
reveal our present ignorance, our advance upon primitive
ideas has been so great that it is difficult for us to realize
what they were without an attentive and not uninstructive

study of them. No other science, not even geology, can
compare with astronomy for the complete revolution which
it has effected in popular notions, or for the change it has
brought about in men's estimate of their place in creation.
It is probable that there will always be men who believe that
the whole universe was made for their benefit; but, however
this may be, we have already learned from astronomy that our
habitation is not that central spot men once deemed it, but
only an ordinary planet circulating round an ordinary star,
just as we are likely also to learn from biology, that we
occupy the position, as animals, of an ordinary family in an
ordinary class.

That we may more perfectly realize this strange revolution
of ideas, we must throw ourselves as far as possible into
the feeling and spirit of our ancestors, when, without the
knowledge we now possess, they contemplated, as they could
not fail to do, the marvellous and awe-inspiring phenomena
of the heavens by night. To them, for many an age, the
sun and moon and stars, with all the planets, seemed absolutely to rise, to shine, and to set; the constellations to
burst out by night in the east, and travel slowly and in
silence to the west; the ocean waves to rise and fall and
beat against the rock-bound shore as if endowed with life;
and even in the infancy of the intellect they must have
longed to pierce the secrets of this mysterious heavenly
vault, and to know the nature of the starry firmament as it

seemed to them, and the condition of the earth which appeared in the centre of these universal movements. The simplest hypothesis was for them the truth, and they believed that the sky was in reality a lofty and extended canopy bestudded with stars, and the earth a vast plain, the solid basis of the universe, on which dwelt man, sole creature that lifted his eyes and thoughts above. Two distinct regions thus appeared to compose the whole system—the upper one, or the air, in which were the moving stars, the lights of heaven, and the firmament over all; and the lower one, or earth and sea, adorned on the surface with the products of life, and below with the minerals, metals, and stones. For a long time the various theories of the universe, grotesque and changing as they might be, were but modifications of this one central idea, the earth below, the heavens above, and on this was based every religious system that was promulgated—the very phrases founded upon it remaining to this day for a testimony to the intimate relation thus manifested between the infant ideas in astronomy and theology. No wonder that early revolutions in the conceptions in one science were thought to militate against the other. It is only when the thoughts on both are enlarged that it is seen that their connection is not necessary, but accidental, or, at least, inevitable only in the infancy of both.

It is scarcely possible to estimate fully the enormous

change from these ideas representing the appearances to
those which now represent the reality; or to picture to
ourselves the total revolution in men's minds before they
could transform the picture of a vast terrestrial surface, to
which the sun and all the heavenly bodies were but acces-
sories for various purposes, to one in which the earth is but
a planet like Mars, moving in appearance among the stars,
as it does, and rotating with a rapidity that brings a whole
hemisphere of the heavens into view through the course of
a single day and night. At first sight, what a loss of dignity!
but, on closer thought, what a gain of grandeur! No longer
some little neighbouring lights shine down upon us from a
solid vault; but we find ourselves launched into the sea of
infinity; with power to gaze into its almost immeasurable
depths.

To appreciate rightly our position, we have to plant ourselves, in imagination, in some spot removed from the surface
of the earth, where we may be uninfluenced by her motion,
and picture to ourselves what we should see. Were we
placed in some spot far enough removed from the earth,
we should find ourselves in eternal day; the sun would ever
shine, for no great globe would interpose itself between it
and our eyes; there would be no night there. Were we in
the neighbourhood of the earth's orbit, and within it, most
wonderful phenomena would present themselves. At one
time the earth would appear but an ordinary planet, smaller

than Venus, but, as time wore on, unmeasured by recurring days or changing seasons, it would gradually be seen to increase in size—now appearing like the moon at the full, and shining like her with a silver light. As it came nearer, and its magnitude increased, the features of the surface would be distinguished; the brighter sea and the darker shining continents, with the brilliant ice-caps at the poles; but, unlike what we see in the moon, these features would appear to move, and, one after another, every part of the earth would be visible. The actual time required for all to pass before us would be what we here call a day and night. And still, as it rotates, the earth passes nearer to us, assumes its largest apparent size, and so gradually decreasing again, becomes once more, after the interval we here call a year, an ordinary-looking star-like planet. To us, in these days, this description is easy of imagination; we find no difficulty in picturing it to ourselves; but, if we will think for a moment what such an idea would have been to the earliest observers of astronomy, we shall better appreciate the vast change that has taken place—how we are removed from them, as we may say, *toto cælo*.

But not only as to the importance of the earth in the universe, but on other matters connected with astronomy, we perceive the immensity of the change in our ideas—in that of distance, for instance. This celestial vault of the ancients was near enough for things to pass from it to us; it was in

close connection with the earth, supported by it, and therefore of less diameter; but now, when our distance from the
sun is expressed by numbers that we may write, indeed,
but must totally fail to adequately appreciate, and the distance from the *next* nearest star is such, that with the velocity
of light—a velocity we are accustomed to regard as instan-
taneous—we should only reach it after a three years' journey,
we are reminded of the pathetic lines of Thomas Hood:

> " I remember, I remember, the fir trees straight and high,
>   And how I thought their slender tops were close against the sky ;
>   It was a childish fantasy, but now 'tis little joy,
>   To know I'm further off from heaven than when I was a boy."

The astronomer's answer to the last line would be that as
far as the material heaven goes, we are just as much in
it as the stars or as any other member of the universe;
we cannot, therefore, be far off or near to it.

It is probable that we are even yet but little awake to
true cosmical ideas in other respects;—as to velocity, for
instance.   We know indeed, of light and electricity and the
motions of the earth, but revelations are now being made to
us of motions of material substances in the sun with such
velocities that in comparison with them any motions on
the earth appear infinitesimally small.   Our progress to
our present notions, and appreciations of the truth of nature
in the heavens, will thus occupy much of our thoughts;
but we must also recount the history of the acquirement

of those facts which have ultimately become the basis for
our changes of idea.

Our rustic forefathers, whatever their nation, were not
so enamoured of the "wonders of science"—that their astronomy was greatly a collection of theories, though theories,
and wild ones, they had; it was a more practical matter,
and was believed too by them to be more practical than
we now find reason to believe to be the case. They noticed
the various seasons, and they marked the changes in the
appearances of the heavens that accompanied them; they
connected the two together, and conceived the latter to be
the cause of the former, and so, with other apparently un
certain events. The celestial phenomena thus acquired a
fictitious importance which rendered their study of primary
necessity, but gave no occasion for a theory.

That we may better appreciate the earliest observations on
astronomy, it may be well to mention briefly what are the
varying phenomena which may most easily be noticed. If
we except the phases of the moon, which almost without
observation would force their recognition on people who
had no other than lunar light by night, and which must
therefore, from the earliest periods of human history have
divided time into lunar months; there are three different
sets of phenomena which depend on the arrangement of our
planetary system, and which were early observed.

The first of these depends upon the earth's rotation on

its axis, the result of which is that the stars appear to revolve with a uniform motion from east to west; the velocity increasing with the distance from the pole star, which remains nearly fixed. This circumstance is almost as easy of observation as the phases of the moon, and was used from the earliest ages to mark the passage of time during the night. The next arises from the motion of the earth in her orbit about the sun, by which it happens that the earth is in a different position with respect to the sun every night, and, therefore, a different set of stars are seen in his neighbourhood; these are setting with him, and therefore also a different set are just rising at sunset every evening. These changes, which would go through the cycle in a year, are, of course, less obvious, but of great importance as marking the approach of the various seasons during ages in which the hour of the sun's rising could not be noted by a clock. The last depends on the proper motions of the moon and planets about the earth and sun respectively, by reason of which those heavenly bodies occupy varying positions among the stars. Only a careful and continuous scrutiny of the heavens would detect these changes, except, perhaps, in the case of the moon, and but little of importance really depends on them; nevertheless, they were very early the subject of observation, as imagination lent them a false value, and in some cases because their connection with eclipses was perceived. The practical cultivation of astronomy amongst the earliest people had always reference to one or other of these three sets of appearances, and the various terms and signs that were invented were intended for the clearer exposition of the results of their observations on these points.

In looking therefore into extreme antiquity we shall find in many instances our only guide to what their knowledge was is the way in which they expressed these results.

We do not find, and perhaps we should scarcely expect to find, any one man or even one nation who laid the foundation of astronomy—for it was an equal necessity for all, and was probably antecedent to the practice of remembering men by their names.   We cannot, either, conjecture the antiquity of ideas and observations met with among races who are themselves the only record of their past ; and if we are to find any origins of the science, it is only amongst those nations which have been cultivators of arts by which their ancient doings are recorded.

Amongst the earliest cultivators of astronomy we may refer to the Primitive Greeks, the Chinese, the Egyptians, the Babylonians, and the Aryans, and also to certain traditions met with amongst many savage as well as less barbarous races, the very universality of which proclaims as loudly as possible their extreme antiquity.

Each of the four above-mentioned races have names with which are associated the beginnings of astronomy—Uranus and Atlas amongst the Greeks ; Folic amongst the Chinese ;

Thaut or Mercury in Egypt; Zoroaster and Bel in Persia and Babylonia. Names such as these, if those of individuals, are not necessarily those of the earliest astronomers—but only the earliest that have come down to us. Indeed it is very far from certain whether these ancient celebrities have any real historical existence. The acts and labours of the earliest investigators are so wrapped in obscurity, there is such a mixture of fable with tradition, that we can have no reliance that any of them, or that others mentioned in ancient mythology, are not far more emblematical than personal. Some, such as Uranus, are certainly symbolical; but the very existence of the name handed down to us, if it prove nothing else, proves that the science was early cultivated amongst those who have preserved or invented them.

If we attempt to name in years the date of the commencement—not of astronomy itself—for that probably in some form was coeval with the race of man itself, but of recorded observations, we are met with a new difficulty arising from the various ways in which they reckoned time. This was in every case by the occurrence of the phases of one or other of the above-mentioned phenomena; sometimes however they selected the apparent rotation of the sun in twenty-four hours, sometimes that of the moon in a month, sometimes the interval from one solstice to the next, and yet they apparently gave to each and all of these the same title—such as *annus*—obviously representing a cycle only, but without reference to its length. By these different methods of counting, hopeless confusion has often been introduced into chronology; and the moderns have in many instances unjustly accused the ancients of vanity and falsehood. Bailly attempted to reconcile all these various methods and consequent dates with each other, and to prove that practical astronomy commenced "about 1,500 years before the Deluge, or that it is about 7,000 years old;" but we shall see reason in the sequel for suspecting any such attempt, and shall endeavour to arrive at more reliable dates from independent evidence.

Perhaps the remotest antiquity to which we can possibly mount is that of the Aryans, amongst whom the hymns of the *Rig Veda* were composed. The short history of Hebrew and Greco-Roman civilization seems to be lost in comparison with this the earliest work of human imagination. When seeking for words to express their thoughts, these primitive men by the banks of the Oxus personified the phenomena of the heavens and earth, the storm, the wind, the rain, the stars and meteors. Here, of course, it is not practical but theoretical astronomy we find. We trace the first figuring of that primitive idea alluded to before—the heaven above, the earth below. Here, as we see, is the earth represented as an indefinite plane surface and passive being forming the foundation of the world; and above it the sky, a luminous and variable vault beneath which shines out the fertile and life-giving light. Thus to the earth they gave the name

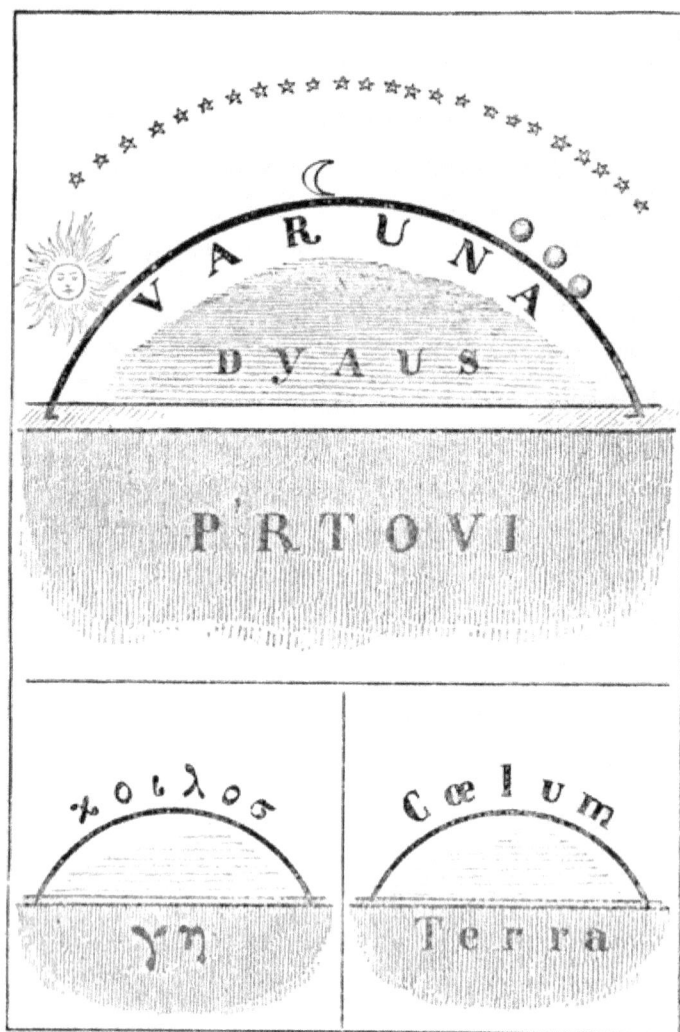

FIG. 1.

P'RTHOVI, "the wide expanse;" the blue and star-bespangled heavens they called VARUNA, "the vault;" and beneath it in the region of the clouds they enthroned the light DYAUS, *i.e.* "the luminous air."

From hence, it would appear, or on this model, the early ideas of all peoples have been formed. Among the Greeks the name for heaven expresses the same idea of a hollow vault (κοῖλος, hollow, concave) and the earth is called γῆ, or mother. Among the Latins the name *cælum* has the same signification, while the earth *terra* comes from the participle *tersa* (the dry element) in contradistinction to *mare* the wet.

In this original Aryan notion, however, as represented by the figure, we have more than this, the origin of the names *Jupiter* and *Deus* comes out. For it is easy to trace the connection between *Dyaus* (the luminiferous air) and the Greek word *Zeus* from whence *Dios*, θεός, *Deus*, and the French word *Dieu*, and then by adding *pater* or father we get *Deuspater*, *Zeuspater*, Jupiter.

These etymologies are not however matters beyond dispute, and there are at least two other modes of deriving the same words. Thus we are told the earliest name for the Deity was Jehovah, the word *Jehov* meaning father of life; and that the Greeks translated this into *Dis* or *Zeus*, a word having, according to this theory, the same sense, being derived from ζαω to live. Of course there can be no question of the later word *Deus* being the direct translation of *Dios*.

A third theory is that there exists in one of the dialects which formed the basis of the old languages of Asia, a word *Yahouh*, a participle of the verb *nih*, to exist, to be; which therefore signifies the self-existent, the principle of life, the origin of all motion, and this is supposed to be the allusion of Diodorus, who explaining the theology of the Greeks, says that the Egyptians according to Manetho, priest of Memphis, in giving names to the five elements have called the spirit or ether Youpiter in the *proper sense* of the word, for the spirit is the source of life, the author of the vital principle in animals, and is hence regarded as the father or generator of all beings. The people of the Homeric ages thought the lightning-bearing Jupiter was the commencement, origin, end, and middle of all things, a single and universal power, governing the heavens, the earth, fire, water, day and night, and all things. Porphyry says that when the philosophers discoursed on the nature and parts of the Deity, they could not imagine any single figure that should represent all his attributes, though they presented him under the appearance of a man, who was *seated* to represent his immovable essence; uncovered in his upper part, because the upper parts of the universe or region of the stars manifest most of his nature; but clothed below the loins, because he is more hidden in terrestrial things; and holding a sceptre in his left hand, because his heart is the ruler of all things. There are, besides, the etymologies which assert that Jupiter is derived from *juvare* to help, meaning

the assisting father ; or again that he is *Dies pater*—the god
of the day—in which case no doubt the sun would be alluded
to.

It appears then that the ancient Aryan scheme, though
possibly supplying us with the origin of one of the widest
spread of our words, is not universally allowed to do so. This
origin, however, appears to derive support from the apparent
occurrence of the original of another well-known ancient classical word in the same scheme, that is Varuna, obviously the
same word as Oὐρανος, and Uranus, signifying the heavens.
Less clearly too perhaps we may trace other such words to the
same source. Thus the Sun, which according to these primitive
conceptions is the husband of the Earth, which it nourishes
and makes fruitful, was called *Savitr* and *Surya*, from which
the passage to the Gothic *Sauil* is within the limits of known
etymological changes, and so comes the Lithuanian *Saull*,
the Cymric *Haul*, the Greek *Heilos*, the Latin *Sol*, and the
English *Solar*. So from their *Nakt*, the destructive, we get
Nux, *Nacht*, *Night*. From *Glu*, the Shining, whence the participle *Glucina*, and so to *Lucina*, *Lucna*, *Luna*, *Lune*.

Turning from the ancient Aryans, whose astronomy we
know only from poems and fables, and so learn but little of
their actual advance in the science of observation, we come to
the Babylonians, concerning whose astronomical acquirements
we have lately been put in possession of valuable evidence by
the tablets obtained by Mr. Smith from Kouyunjik, an account

of the contents of which has been given by Mr. Sayce (*Nature*, vol. xii. p. 489). As the knowledge thus obtained is more certain, being derived from their actual records, than any that we previously possessed, it will be well to give as full an account of it as we are able.

The originators of Babylonian astronomy were not the Chaldaeans, but another race from the mountains of Elam, who are generally called Acadians. Of the astronomy of this race we have no complete records, but can only judge of their progress by the words and names left by them to the science, as afterwards cultivated by the Semitic Babylonians. These last were a subsequent race, who entering the country from the East, conquered the original inhabitants about 2000 B.C., and borrowed their civilization, and with it their language in the arts and sciences. But even this latter race is one of considerable antiquity, and when we see, as we shortly shall, the great advances they had made in observations of the sun and moon, and consider the probable slowness of development in those early ages, we have some idea of the remoteness of the date at which astronomical science was there commenced. Our chief source of information is an extremely ancient work called *The Observations of Bell*, supposed to have been written before 1700 B.C., which was compiled for a certain King Saigou, of Agave in Babylonia. This work is in seventy books or parts, and is composed of numerous small earthen tablets having impressed upon them the cuneiform character in which

they printed, and which we are now able to read. We
generally date the art of printing from Caxton, in 1474,
because it took the place of manuscript that had been previously in use in the West; but that method of writing, if in
some respects an improvement on previous methods of recording ideas as more easily executed, was in others a retrogression as being less durable: while the manuscripts have perished
the impressions on stone have remained to this day, and will
no doubt last longer than even our printed books. These
little tablets represented so many leaves, and in large librar-
ies, such as that from which those known have been derived,
they were numbered as our own are now, so that any particular one could be asked for by those who might wish to
consult it. The great difficulty of interpreting these records,
which are written in two different dialects, and deal often with
very technical matters, may well be imagined. These difficulties however have been overcome, and a good approach to the
knowledge of their contents has been made. The Chaldæans,
as is well known, were much given to astronomy and
many of their writings deal with this subject; but they did
practical work as well, and did not indulge so much in theory
as the Aryans. We shall have future occasion in this book
to refer to their observations on various points, as they did
not by any means confine themselves to the simplest matters;
much, in fact, of that with which modern astronomy deals,
the dates and duration of eclipses of the sun and moon, the

accurate measurement of time, the existence of cycles in
lunar and solar phenomena, was studied and recorded by them.
We can make some approach to the probable dates of the
invention of some part of their system, by means of the
signs of the Zodiac, which were invented by them and which
we will discuss more at length hereafter. We need only
say at present that what is now the sign of spring, was
not reckoned so with them, and that we can calculate how
long ago it is that the sign they reckoned the spring sign
was so.

Semiramis also raised in the centre of Babylon a temple
consecrated to Jupiter, whom the Babylonians called Bel.
It was of an extraordinary height and served for an observa-.
tory. The whole edifice was constructed with great art in
asphalte and brick. On its summit were placed the statues
of Jupiter, Juno, and Rhea, covered with gold.

The Egyptians have always been named as the earliest
cultivators of astronomy by the Grecian writers, by whom the
science has been handed down to us, and the Chaldæans have
even been said to have borrowed from them. The testimony
of such writers however is not to be received implicitly, but
to be weighed with the knowledge we may now obtain, as
we have noticed above with respect to the Babylonians, from
the actual records they have left us, whether by actual
records, or by words and customs remaining to the present
day.

PLATE I.—BABYLONIAN ASTRONOMERS.

Herodotus declares that the Egyptians had made observations for 11,340 years and had seen the course of the sun
change four times, and the ecliptic placed perpendicular to the
equator. This is the style of statement on which opinions
of the antiquity of Egyptian astronomy have been founded,
and it is obviously unworthy of credit.

Diodorus says that there is no country in which the
positions and motions of the stars have been so accurately
observed as in Egypt (i.e. to his knowledge). They have
preserved, he says, for a great number of years registers in
which their observations are recorded. Expositions are found
in these registers of the motions of the planets, their revolutions and their stations, and, moreover, the relation which
each bears to the birthdays of animals, and its good or evil
influence. They often predicted the future with success.
The earthquakes, inundations, the appearance of comets, and
many other phenomena which it is impossible for the vulgar
to know beforehand, were foreseen by them by means of the
observations they had made over a long series of years.

On the occasion of the French expedition to Egypt, a long
passage was discovered leading from Karnak to Lu_ksor.
This passage was adorned on each side of the way with a range
of 1600 sphinxes with the body of a lion and the head of a
ram. Now in Egyptian architecture, the ornaments are never
the result of caprice or chance; on the contrary, all is done
with intention, and what often appears at first sight strange,

appears, after having been carefully examined and studied, to present allegories full of sense and reason, founded on a profound knowledge of natural phenomena, that the ornaments are intended to record.   These sphinxes and rams of the passage were probably the emblems of the different signs of the Zodiac along the route of the sun.   The date of the avenue is not known; but it would doubtless lead us to a high antiquity for the Egyptian observations.

The like may be said of the great pyramid, which according to Piazzi Smyth was built about 2170 B.C.   Certainly there are no carvings about it exhibiting any astronomical designs ; but the exact way in which it is executed would seem to indicate that the builders had a very clear conception of the importance of the meridian line.   It should, however, be stated that Piazzi Smyth does not consider it to have been built by the Egyptians for themselves ; but under the command of some older race.

There seem, however, to be indications in various festivals and observances, which are met with widely over the earth's surface, as will be indicated more in detail in the chapter on the Pleiades, that some astronomical observations, though of the rudest, were made by races anterior even to those whose history we partially possess ; and that not merely because of its naturalness, but because of positive evidence, we must trace back astronomy to a source from whence Egyptians, Indians, and perhaps Babylonians themselves derived it.

The Chinese astronomy is totally removed from these and stands on its own basis. With them it was a matter concerning the government, and stringent laws were enforced on the state astronomers. The advance, however, that they made would appear to be small; but if we are to believe their writers, they made observations nearly three thousand years before our era.

Under the reign cf Hoangti, Yuchi recorded that there was a large star near the poles of the heavens. By a method which we shall enlarge upon further on, it can be astronomically ascertained that about the epoch this observation was said to be made there was a star (a Draconis) so near the pole as to appear immovable, which is so far a confirmation of his statement. In 2169 the first of a series of eclipses was recorded by them; but the value of their astronomy seems to be doubtful when we learn that calculation proves that not one of them previous to the age of Ptolemy can be identified with the dates given.

Amongst all nations except the Chinese, where it was political, and the Greeks, where it was purely speculative, astronomy has been intimately mixed with religious ideas, and we consequently find it to have taken considerable hold on the mind.

Just as we have seen among the Indians that the basis of their astronomical ideas was the two-fold division into heaven and earth, so among other nations this duality has

formed the basis of their religion. Two aspects of things
have been noticed by men in the constitution of things—that
which remains always, and that which is merely transitory,
causes and effects. The heaven and the earth have presented
the image of this to their minds—one being the eternal existence, the other the passing form. In heaven nothing seems
to be born, increase, decrease, or die above the sphere of the
moon. That alone showed the traces of alteration in its
phases; while on the other hand there was an image of
perpetuity in its proper substance, in its motion, and the
invariable succession of the same phases.

From another point of view, the heavens were regarded as
the father, and the earth-as the mother of all things. For the
principle of fertility in the rains, the dew and the warmth,
came from above; while the earth brought forth abundantly
of the products of nature. Such is the idea of Plutarch, of
Hesiod, and of Virgil. From hence have arisen the fictions
which have formed the basis of theogony. Uranus is said to
have espoused Ghe, or the heavens took the earth to wife, and
from their marriage was born the god of time or Saturn.

Another partly religious, and partly astronomical antago-
nism has been drawn between light and darkness, associated
respectively with good and evil. In the days when artificial
lights, beyond those of the flickering fire, were unknown, and
with the setting of the sun all the world was enveloped in
darkness and seemed for a time to be without life, or at least

cut off entirely from man, it would seem that the sun and
its light was the entire origin of life. Hence it naturally
became the earliest divinity whose brilliant light leaping out
of the bosom of chaos, had brought with it man and all the
universe, as we see it represented in the theologies of Orpheus
and of Moses; whence the god Bel of the Chaldeans, the
Oromaza of the Persians, whom they invoke as the source of
all that is good in nature, while they place the origin of all
evil in darkness and its god Ahrinam. We find the glories
of the sun celebrated by all the poets, and painted and represented by numerous emblems and different names by the
artists and sculptors who have adorned the temples raised to
nature or the great first cause.

Among the Jews there are traditions of a very high antiquity for their astronomy. Josephus assures us that it was
cultivated before the Mosaic Deluge. According to him it is
to the public spirit and the labour of the antediluvians that we
owe the science of astrology: " and since they had learnt from
Adam that the world should perish by water and by fire,
the fear that their science should be lost, made them erect two
columns, one of brick the other of stone, on which they engraved the knowledge they had acquired, so that if a deluge
should wash away the column of brick, the stone one might
remain to preserve for posterity the memory of what they
had written. The prescience was rewarded, and the column
of stone is still to be seen in Syria." Whatever we may think

of this statement it would certainly be interesting if we could find in Syria or anywhere else a monument that recorded the ancient astronomical observations of the Jews. Ricard and others believe that they were very far advanced in the science, and that we owe a great part of our present astronomy to them; but such a conjecture must remain without proof unless we could prove them anterior to the other nations, whom, we have seen, cultivated astronomy in very remote times.

One observation seems peculiar to them, if indeed it be a veritable observation. Josephus says, "God prolonged the life of the patriarchs that preceded the deluge, both on account of their virtues, and to give them the opportunity of perfecting the sciences of geometry and astronomy which they had discovered; which they could not have done if they had not lived for 600 years, because it is only after the lapse of 600 years that the *great year* is accomplished."

Now what is this great year or cycle of 600 years? M. Cassini, the director of the Observatory of Paris, has discussed it astronomically. He considers it as a testimony of the high antiquity of their astronomy. "This period," he says, "is one of the most remarkable that have been discovered; for, if we take the lunar month to be 29 days 12h. 44m. 3s. we find that 219,146½ days make 7,421 lunar months, and that this number of days gives 600 solar years of 365 days 5h. 51m. 36s. If this year was in use before the deluge, it appears very probable it must be acknowledged that the patriarchs were already acquainted to a considerable degree of accuracy with the motions of the stars, for this lunar month agrees to a second almost with that which has been determined by modern astronomers."

A very similar argument has been used by Prof. Piazzi Smyth to prove that the Great Pyramids were built by the descendants of Abraham near the time of Noah; namely, that measures of two different elements in the measurement of time or space when multiplied or divided produce a number which may be found to represent some proportion of the edifice, and hence to assume that the two numbers were known to the builders.

We need scarcely point out that numbers have always been capable of great manipulation, and the mere fact of one number being so much greater than another, is no proof that *both* were known, unless we knew that *one* of them was known independently, or that they are intimately connected.

In the case of Josephus' number the cycle during which the lunar months and solar years are commensurable has been long discussed and if the number had been 19 instead of 600, we should have had little doubt of its reference; yet 600 is a very simple number and might refer to many other cycles than the complicated one pointed out by M. Cassini. A similar case may be quoted with regard to the Indians, which, according to our temperament, may be either considered a proof that these reasonings are correct, or that they are easy to make. They say that there are two stars diametrically opposite which pass through the zodiac in 144 years; nothing can be made of this period, nor yet of another equally problematical one of 180 years; but if we multiply the two together we obtain 25,9×20, which is very nearly the length of the cycle for the precession of the equinoxes.

In this review of the ancient ideas of different peoples, we have followed the most probable order in considering that the observation of nature came first, and the different parts of it were afterwards individualized and named. It is proper to add that according to some ancient authors— such as Diodorus Siculus—the process was considered to have been the other way. That Uranus was an actual individual, that Atlas and Saturn were his sons or descendants or followers, and that because Atlas was a great astronomer he was said to support the heavens, and that his seven daughters were real, and being very spiritual they were regarded as goddesses after death and placed in heaven under the name of the Pleiades.

However, the universality of the ideas seems to forbid this interpretation, which is also in itself much less natural.

These various opinions lead us to remark, in conclusion,

that the fables of ancient mythological astronomy must be
interpreted by means of various keys. Allegory is the
first—the allegory employed by philosophers and poets
who have spoken in figurative language. Their words taken
in the letter are quite unnatural, but many of the fables
are simply the description or explanation of physical facts.
Hieroglyphics are another key. Having become obscure by
the lapse of time they sometimes, however, present ideas
different from those which they originally expressed. It is
pretty certain that hieroglyphics have been the source of
the men with dogs' heads, or feet of goats, &c. Fables also
arise from the adoption of strange words whose sound is
something like another word in the borrowing language connected with other ideas, and the connection between the
two has to be made by fable.

CHAPTER II.

THE numerous stone monuments that are to be found scattered over this country, and over the neighbouring parts of Normandy, have given rise to many controversies as to their origin and use. By some they have been supposed to be mere sepulchral monuments erected in late times since the Roman occupation of Great Britain. Such an idea has little to rest upon, and we prefer to regard them, as they have always been regarded, as relics of the Druidical worship of the Celtic or Gaulish races that preceded us in this part of Europe.

If we were to believe the accounts of ordinary historians, we might believe that the Druids were nothing more than a kind of savage race, hidden, like the fallow-deer in the recesses of their woods. Thought to be sanguinary, brutal, superstitious, we have learned nothing of them beyond their human sacrifices, their worship of the oak, their raised stones ; without inquiring whether these characteristics which scandalize our tastes, are not simply the legacy of a primitive era, to which, by the side of the tattered religions of the old Paganism, Druidism remained faithful. Nevertheless the Druids were not without merit in the order of thought.

For the Celts, as for all primitive people, astronomy and religion were intimately associated. They considered that the soul was eternal, and the stars were worlds successively inhabited by the spiritual emigrants. They considered that the stars were as much the abodes of human life as our own earth, and this image of the future life constituted their power and their grandeur. They repelled entirely the idea of the destruction of life, and preferred to see in the phenomena of death, a voyage to a region already peopled by friends.

Under what form did Druidical science represent the universe ? Their scientific contemplation of the heavens was at the same time a religious contemplation. It is therefore impossible to separate in our history their astronomical and theological heavens.

In their theological astronomy, or astronomical theology, the Druids considered the totality of all living beings as divided into three circles. The first of these circles, the circle of immensity, *Ceugant*, corresponding to incommunicable, infinite attributes, belonged to God alone; it was properly the absolute, and none, save the ineffable being, had a right there. The second circle, that of blessedness, *Gwyn-fyd*, united in it the beings that have arrived at the superior degrees of existence; this was heaven. The third, the circle of voyages, *Abred*, comprised all the noviciate; it was there, at the bottom of the abysses, in the great oceans, as Taliesin says, that the first breath of man commenced. The object proposed to men's perseverance and courage was to attain to what the bards called the point of liberty, very probably the point at which, being suitably fortified against the assaults of the lower passions, they were not exposed to be troubled, against their wills, in their celestial aspirations; and when they arrived at such a point—so worthy of the ambition of every soul that would be its own master—they quitted the circle of Abred and entered that of Gwyn-fyd; the hour of their recompense had come.

Demetrius, cited by Plutarch, relates that the Druids believed that these souls of the elect were so intimately connected with our circle that they could not emerge from it without disturbing its equilibrium. This writer states, that being in the suite of the Emperor Claudius, in some part of the British isles, he heard suddenly a terrible hurricane, and the priests, who alone inhabited these sacred islands, immediately explained the phenomenon, by telling him that a vacuum had been produced on the earth, by the departure of an important soul. "The great men," he said,

" while they live are like torches whose light is always beneficent and never harms any one, but when they are extinguished their death generally occasions, as you have just seen, winds, storm, and derangements of the atmosphere."

The palingenetic system of the Druids is complete in itself, and takes the being at his origin, and conducts him to the ultimate heaven. At the moment of his creation, as Henry Martyn says in his Commentary, the being has no conscience of the gifts that are latent in him. He is created in the lowest stage of life, in *Annwfn*, the shadowy abyss at the base of *Abred*. There, surrounded by nature, submitted to necessity, he rises obscurely through the successive degrees of inorganic matter, and then through the organic. His conscience at last awakes. He is man. " Three things are primarily contemporaneous—man, liberty, and light." Before man there was nothing in creation but fatal obedience to physical laws; with man commences the great battle between liberty and necessity, good and evil. The good and the evil present themselves to man in equilibrium, " and he can at his pleasure attach himself to one or the other of them."

It might appear at first sight that it was carrying things too far to attribute to the Druids the knowledge, not indeed of the true system of the world, but the general idea on which it was constructed. But, on closer examination, this

opinion seems to have some consistency. If it was from
the Druids that Pythagoras derived the basis of his theology,
why should it not be from them that he derived also that
of his astronomy? Why, if there is no difficulty in seeing
that the principle of the subordination of the earth might
arise from the meditations of an isolated spirit, should
there be any more difficulty in thinking that the principles
of astronomy should take birth in the midst of a corporation
of theologians embued with the same ideas as the philosophers on the circulation of life, and applied with continued
diligence to the study of celestial phenomena. The Druid,
not having to receive mythological errors, might be led by
that circumstance to imagine in space other worlds similar
to our own.

Independently of its intrinsic value, this supposition rests
also upon the testimony of historians. A singular statement
made by Hecatæus with regard to the religious rites of Great
Britain exhibits this in a striking manner. This historian
relates that the moon, seen in this island, appears much
larger than it does anywhere else, and that it is possible to
distinguish mountains on its surface, such as there are on
the earth. Now, how had the Druids made an observation
of this kind? It is of not much consequence whether they
had actually seen the lunar mountains or had only imagined
them, the curious thing is that they were persuaded that
that body was like the earth, and had mountains and other

features similar to our own. Plutarch, in his treatise *De facie in orbe Lunæ*, tells us that, according to the Druids, and conformably to an idea which had long been held in science, the surface of the moon is furrowed with several Mediterraneans, which the Grecian philosophers compare to the Red and Caspian seas. It was also thought that immense abysses were seen, which were supposed to be in communication with the hemisphere that is turned away from the earth. Lastly, the dimensions of this sky-borne country were estimated; (ideas very different to those that were current in Greece): its size and its breadth, says the traveller depicted by the writer, are not at all such as the geometers say, but much larger.

It is through the same author, who is in accordance in this respect with all the bards, that we know that this celestial earth was considered by the theologians of the West as the residence of happy souls. They rose and approached it in proportion as their preparation had been complete, but, in the agitation of the whirlwind, many reached the moon that it would not receive. "The moon repelled a great number, and rejected them by its fluctuations, at the moment they reached it; but those that had better success fixed themselves there for good; their soul is like the flame, which, raising itself in the ether of the moon, as fire raises itself on that of the earth receives force and solidity in the same way that red-hot iron does when plunged into the water."

They thus traced an analogy between the moon and the earth, which they doubtless carried out to its full development, and made the moon an image of what they knew here, picturing there the lunar fields and brooks and breezes and perfumes. What a charm such a belief must have given to the heavens at night. The moon was the place and visible pledge of immortality. On this account it was placed in high position in their religion : the order of all the festivals was arranged after that which was dedicated to it ; its presence was sought in all their ceremonies, and its rays were invoked. The Druids are always therefore represented as having the crescent in their hands.

Astronomy and theology being so intimately connected in the spirit of the Druids, we can easily understand that the two studies were brought to the front together in their colleges. From certain points of view we may say that the Druids were nothing more than astronomers. This quality was not less striking to the ancients in them than in the Chaldæans. The observation of the stars was one of their official functions. Cæsar tells us, without entering more into particulars, that they taught many things about *the form and dimensions of the earth, the size and arrangements of the different parts of heaven, and the motions of the stars*, which includes the greater part of the essential problems of celestial geometry, which we see they had already proposed to themselves. We can see the same fact in the magnificent passage of Taliesin. "I will ask the bards," he says in his *Hymn of the World*, "and why will not the bards answer me? I will ask of them what sustains the earth, since having no support it does not fall? or if it falls which way does it go? But what can serve for its support? Is the world a great traveller? Although it moves without ceasing, it remains tranquil in its route; and how admirable is that route, seeing that the world moves not in any direction." This suffices to show that the ideas of the Druids on material phenomena were not at all inferior to their conceptions of the destiny of the soul, and that they had scientific views of quite another origin from the Alexandrian Greeks, the Latins, their disciples, or the middle ages. An anecdote of the eighth century furnishes another proof in favour of Druidical science. Every one knows that Virgilius, bishop of Salzburg, was accused of heresy by Boniface before the Pope Zacharias, because he had asserted that there were antipodes. Now Virgilius was educated in one of the learned monasteries of Ireland, which were fed by the Christian bards, who had preserved the scientific traditions of Druidism.

The fundamental alliance between the doctrine of the plurality of worlds and of the eternity of the soul is perhaps the most memorable character in the thoughts of this ancient race. The death upon earth was for them only a psycho-logical and astronomical fact, not more grave than that which

PLATE II.—DRUIDICAL WORSHIP.

happened to the moon when it was eclipsed, nor the fall of the verdant clothing of the oak under the breath of the autumnal breeze. We see these conceptions and manners, at first sight so extraordinary, clothe themselves with a simple and natural aspect. The Druids were so convinced of the future life in the stars, that they used *to lend money to be repaid in the other world.* Such a custom must have made a profound impression on the minds of those who daily practised it. Pomponius Mela and Valerius Maximus both tell us of this custom. The latter says, "After having left Marseilles I found that ancient custom of the Gauls still in force, namely, of lending one another money to be paid back in the infernal regions, for they are persuaded that the souls of men are immortal."

In passing to the other world they lost neither their personality, their memory, nor their friends; they there re-encountered the business, the laws, the magistrates of this world. They had capitals and everything the same as here. They gave one another rendezvous as emigrants might who were going to America. This superstition, so laudable as far as it had the effect of pressing on the minds of men the firm sentiment of immortality, led them to burn, along with the dead, all the objects which had been dear to them, or of which they thought they might still wish to make use. "The Gauls," says Pomponius Mela, "burn and bury with the dead that which had belonged to the living."

They had another custom prompted by the same spirit,
but far more touching. When any one bade farewell to the
earth, each one charged him to take letters to his absent
friends, who should receive him on his arrival and doubtless
load him with questions as to things below. It is to Diodorus that we owe the preservation of the remembrance
of this custom. "At their funerals," he says, "they place
letters with the dead which are written to those already dead
by their parents, so that they may be read by them." They
followed the soul in thought in its passage to the other
planets, and the survivors often regretted that they could
not accomplish the voyage in their company; sometimes,
indeed, they could not resist the temptation. "There are
some," says Mela, "who burn themselves with their friends
in order that they may continue to live together." They
entertained another idea also, which led even to worse practices than this, namely, that death was a sort of recruiting
that was commanded by the laws of the universe for the
sustenance of the army of existences. In certain cases they
would replace one death by another. Posidonius, who visited
Gaul at an epoch when it had not been broken up, and who
knew it far better than Cæsar, has left us some very curious
information on this subject. If a man felt himself seriously
warned by his disease that he must hold himself in readiness
for departure, but who, nevertheless, had, for the moment,
some important business on hand, or the needs of his family

chained him to this life, or even that death was disagreeable to him; if no member of his family or his clients were willing to offer himself instead, he looked out for a substitute; such a one would soon arrive accompanied by a troop of friends, and stipulating for his price a certain sum of money, he distributed it himself as remembrances among his companions,—often even he would only ask for a barrel of wine. Then they would erect a stage, improvise a sort of festival, and finally, after the banquet was over, our hero would lie down on the shield, and driving a sword into his bosom, would take his departure for the other world.

Such a custom, indeed, shows anything but what we should rightly call civilization, however admirable may have been their opinions; but it receives its only palliation from the fact that their indifference to death did not arise from their undervaluing life here, but that they had so firm a belief in the existence and the happiness of a life hereafter.

That these beliefs were not separated from their astronomical ideas is seen from the fact that they peopled the firmament with the departed. The Milky Way was called the town of Gwyon (Coër or Ker Gwydion, Ker in Breton, Caer in Gaulish, Kohair in Gaelic); certain bardic legends gave to Gwyon as father a genius called Don, who resides in the constellation of Cassiopeia, and who figures as "the king of the fairies" in the popular myths of Ireland. The empyrean is thus divided between various heavenly spirits. Arthur had for residence the Great Bear, called by the Druids "Arthur's Chariot."

We are not, however, entirely limited to tradition and the reports of former travellers for our information as to the astronomy of the Druids, but we have also at our service numerous coins belonging to the old Gauls, who were of one family with those who cultivated Druidism in our island, which have been discovered buried in the soil of France. The importance which was given to astronomy in that race becomes immediately evident upon the discovery of the fact that these coins are marked with figures having reference to the heavenly bodies, in other words are astronomical coins. If we examine, from a general point of view, a large collection of Gaulish medals such as that preserved in the National Museum of Paris, we observe that among the essential symbols that occupy the fields are types of the Horse, the Bull, the Boar, the Eagle, the Lion, the Horseman, and the Bear. We remark next a great number of signs, most often astronomical, ordinarily accessory, but occasionally the chief, such as the sign ℺, globules surrounded by concentric circles, stars of five, six, or eight points, radiated and flaming bodies, crescents, triangles, wheels with four spokes, the sign ∞, the lunar crescent, the zigzag, &c. Lastly, we remark other accessory types represented by images of real objects or imaginary figures, such as the Lyre, the Diota, the Serpent, the Hatchet, the

Human Eye, the Sword, the Bough, the Lamp, the Jewel, the Bird, the Arrow, the Ear of Corn, the Fishes, &c.

On a great number of medals, on the stateres of Verein-getorix, on the reverses of the coins of several epochs, we recognize principally the sign of the Waterer, which appears to symbolize for one part of antiquity the knowledge of the heavenly sphere. On the Gaulish types this sign (an amphora with two handles) bears the name of Diota, and represents amongst the Druids as amongst the Magi the sciences of astronomy and astrology.

Some of these coins are represented in the woodcut below.

FIG. 2.

The first of these represents the course of the Sun-Horse reaching the Tropic of Cancer (summer solstice), and brought back to the Tropic of Capricorn (winter solstice).

On the second is seen the symbol of the year between the south (represented by the sun ☉) and the north (represented by the Northern Bear). In the third the calendar (or course of the year) between the sun ☉ and the moon ☾. Time the Sun, and the Bear are visible on the fourth. The diurnal motion of the heavens is represented on the fifth; and lastly, on the sixth, appears the Watering-pot, the Sun-Horse, and the sign of the course of the heavenly bodies.

On other groups of money the presence of the zodiac may be made out.

These medals would seem to show that some part of the astronomical knowledge of the Druids was not invented by themselves, but borrowed from the Chaldeans or others who in other lands invented them in previous ages, and from whom they may have possibly derived them from the Phenicians.

We may certainly expect, however, from these pieces of money, if found in sufficient number and carefully studied, to discover a good many positive facts now wanting to us, of the religion, sciences, manners, language, commercial relation, &c. which belonged to the Celtic civilization. It was far from being so barbarous as is ordinarily supposed, and we shall do more justice to it when we know it better.

M. Fillioux, the curator of the museum of Guéret, who has studied these coins with care, after having sought for a long time for a clear and concise method of determining exactly the symbolic and religious character of the Gaulish money, has been able to give the following general statements.

The coins have for their ordinary field the heavens.

On the right side they present almost universally the ideal heads of gods or goddesses, or in default of these, the symbols that are representative of them.

On the reverse for the most part, they reproduce, either by direct types or by emblems artfully combined, the principal celestial bodies, the divers aspects of the constellations, and probably the laws, which, according to their ancient science, presided over their course; in a smaller proportion they denote the religious myths which form the base of the national belief of the Gauls. As we have seen above, for them the present life was but a transitory state of the soul, only a prodrome of the future life, which should develop itself in heaven and the astronomical worlds with which it is filled.

Borrowed from an elevated spiritualism, incessantly tending towards the celestial worlds, these ideas were singularly appropriate to a nation at once warlike and commercial. These circumstances explain the existence of these strange types, founded at the same time on those of other nations, and on the symbolism which was the soul of the Druidical religion. To this religious caste, indeed, we must give the merit of this ingenious and original conception, of turning the reverses of the coins into regular charts of the heavens. Nothing indeed could be better calculated to inspire the people with respect and confidence than these mysterious and learned symbols, representing the phenomena of the heavens.

Not making use of writing to teach their dogmas, which they wished to maintain as part of the mysteries of their caste, the Druids availed themselves of this method of placing on the money that celestial symbolism of which they alone possessed the key.

The religious ideas founded on astronomical observations were not peculiar to, or originated by, the Druids, any more than their zodiac. There seems reason to believe that they had come down from a remote antiquity, and been widely spread over many nations, as we shall see in the chapter on the Pleiades; but we can certainly trace them to the East, where they first prevailed in Persia and Egypt, and were afterwards brought to Greece, where they disappeared before the new creations of anthropomorphism, though they were not forgotten in the days of the poet Anacreon, who says, "Do not represent for me, around this vase" (a vase he had ordered of the worker in silver), "either the heavenly bodies, or the chariot, or the melancholy Orion; I have nothing to do with the Pleiades or the Herdsman." He only wanted mythological subjects which were more to his taste.

The characters which are made use of in these astro-

nomical moneys of the Druids would appear to have a more
ancient origin than we are able to trace directly, since they
are most of them found on the arms and implements of the
bronze age. Some of them, such as the concentric pointed
circles, the crescent with a globule or a star, the line in
zigzag, were used in Egypt; where they served to mark the
sun, the month, the year, the fluid element; and they
appear to have had among the Druids the same signification.
The other signs, such as the $\sim$, and its multiple com-
binations, the centred circles, grouped in one or two, the
little rings, the alphabetical characters recalling the form
of a constellation, the wheel with rays, the radiating discs,
&c. are all represented on the bronze arms found in the
Celtic, Germanic, Breton, and Scandinavian lands. From
this remote period, which was strongly impressed with the
Oriental genius, we must date the origin of the Celtic symbolism. It has been supposed, and not without reason, that
this epoch, besides being contemporaneous with the Phenician establishments on the borders of the ocean, was an age
of civilization and progress in Gaul, and that the ideas of
the Druids became modified at the same time that they
acquired just notions in astronomy and in the art of casting
metals. At a far later period, the Druidic theocracy having,
with religious care, preserved the symbols of its ancient
traditions, had them stamped on the coins which they
caused to be struck.

This remarkable fact is shown in an incontestable manner in the rougher attempts in Gaulish money, and this same state of things was perpetuated even into the epoch of the high arts, since we find on the imitation statues of Macedonia the old Celtic symbols associated with emblems of a Grecian origin.

In Italy a different result was arrived at, because the warlike element of the nobles soon predominated over the religious. Nevertheless the most ancient Roman coins, those which are known to us under the name of Consular, have not escaped the common law which seems to have presided, among all nations, over the origin of money. The two commonest types, one in bronze of *Janus Bifrons* with the *palus;* the other in silver, the *Dioscures* with their stars, have an eminently astronomical aspect.

The comparison between the Gaulish and Roman coins may be followed in a series of analogies which are very remarkable from an astronomical point of view. To cite only a few examples, we may observe on a large number of pennies of different families, the impression of Auriga "the Coachman" conducting a quadriga; or the sun under another form (with his head radiated and drawn in profile); or Diana with her lunar attributes; or the five planets well characterised; for example, Venus by a double star, as that of the morning or of the evening; or the constellations of the Dog, Hercules, the Kid, the Lyre, and almost all

those of the zodiac and of the circumpolar region and
the seven-kine (septemtriones). In later times, under the
Cæsars, in the villa of Borghèse, is found a calendar whose
arrangements very much recall the ancient Gaulish coin.
The head of the twelve great gods and the twelve signs of
the zodiac are represented, and the drawing of the constellations establishes a correspondence between their rising
and the position of the sun in the zodiac. It may therefore
be affirmed that in the coinage and works of art in Italy
and Greece, the characteristic influence of astronomical
worship is found as strongly as among the Druids. Nor
have the Western nations alone had the curious habit of
impressing their astronomical ideas upon their coinage, for
in China and Japan coins of a similar description have been
met with, containing on their reverse all the signs of the
zodiac admitted by them.

In conclusion, we may say, that it was cosmography, that
constructed the dogmas of the Druidical religion, which was,
in its essential elements, the same as that of the old Oriental
theocracies. The outward ceremonies were addressed to the
sun, the moon, the stars, and other visible phenomena; but,
above nature, there was the great generating and moving
principle, which the Celts placed, at a later period perhaps,
among the attributes of their supreme deities.

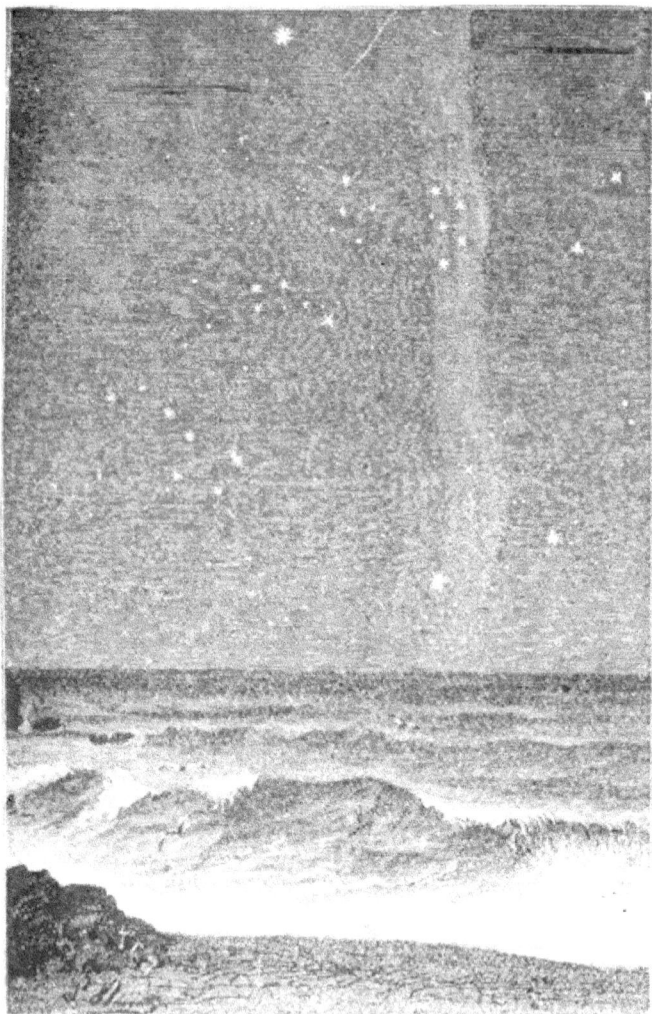

THE NORTHERN CONSTELLATIONS.

The Lyre—Cassiopeia—The Little Bear—The Dragon—Andromeda
—The Great Bear—Capella—Algol, or Medusa's Head.

CHAPTER III.

WHEN we look upon the multitude of heavenly bodies with which the celestial vault is strewed, our attention is naturally arrested by cértain groupings of brilliant stars, apparently associated together on account of their great proximity ; and also by certain remarkable single stars which have excessive brilliancy or are completely isolated from the rest. These natural groups seem to have some obscure connection with or dependence on each other. They have always been noticed, even by the most savage races. The languages of several such races contain different names for the same identical groups, and these names, mostly borrowed from terrestrial beings, give an imaginary life to the solitude and silence of the skies. A celestial globe, as we know, presents us with a singular menagerie, rich in curious monsters placed in inconceivable positions. How these constellations, as they are called, were first invented,

E

and by whom, is an interesting question which by the
aid of comparative philology we must endeavour now to
answer.

Among these constellations there are twelve which have
a more than ordinary importance, and to which more
attention has always been paid. They are those through
which the sun appears to pass in his annual journey round
the ecliptic, entering one region each month. At least, this
is what they were when first invented. They were called
the zodiacal constellations or signs of the zodiac—the name
being derived from their being mostly named after living
beasts. In our own days the zodiacal constellations are no
longer the signs of the zodiac. When they were arranged
the sun entered each one on a certain date. He now is no
longer at the same point in the heavens at that date, nevertheless he is still said to enter the same sign of the zodiac—
which therefore no longer coincides with the zodiacal constellation it was named from—but merely stands for a
certain twelfth part of the ecliptic, which varies from time
to time. It will be of course of great interest to discover
the origin of these particular constellations, the date of
their invention, &c.; and we shall hope to do so after
having discussed the origin of those seen in the Northern
hemisphere which may be more familiar even than those.

We have represented in the frontispiece the two halves
of the Grecian celestial sphere—the Northern and the

Southern, with the various constellations they contain. This sphere was not invented by the Greeks, but was received by them from more ancient peoples, and corrected and augmented. It was used by Hipparchus two thousand years ago; and Ptolemy has given us a description of it. It contained 48 constellations, of which 21 belonged to the Northern, 15 to the Southern hemisphere, and the remaining twelve were those of the zodiac, situated along the ecliptic.

The constellations reckoned by Ptolemy contained altogether 1,026 stars, whose relative positions were determined by Hipparchus; with reference to which accomplishment Pliny says, " Hipparchus, with a height of audacity too great even for a god, has ventured to transmit to posterity the number of the stars!"

Ptolemy's catalogue contains :—

For the northern constellations .	361	stars
For the zodiacal	350	,,
For the southern	318	.,
or	———	
For all the 48 constellations .	1,029	,,
or, since 3 of these are named twice	1,026	,,

Of course this number is not to be supposed to represent the whole of the stars visible even to the naked eye; there are twice as many in the Northern hemisphere alone, while there are about 5,000 in the whole sky. The number visible in a telescope completely dwarfs this, so that more than 300,000 are now catalogued; while the number visible in a large telescope may be reckoned at not less than 77 millions. The principal northern constellations named by Ptolemy are contained in the following list, with the stars of the first magnitude that occur in each:—

The Great Bear, or David's Chariot, near the centre.

The Little Bear, with the Pole Star at the end of the tail.

The Dragon.

Cepheus, situated to the right of the Pole.

The Herdsman, or the Keeper of the Bear, with the star Arcturus.

The Northern Crown to the right.

Hercules, or the Man who Kneels.

The Lyre, or Falling Vulture, with the beautiful star Vega.

The Swan, or Bird, or Cross.

Cassiopeia, or the Chair, or the Throne.

Perseus.

The Carter, or the Charioteer, with Capella Ophiuchus, or Serpentarius, or Esculapius.

The Serpent.

The Bow and Arrow, or the Dart.

The Eagle, or the Flying Vulture, with Altaïr.

The Dolphin.

The Little Horse, or the Bust of the Horse.

Pegasus, or the Winged Horse, or the Great Cross.

Andromeda, or the Woman with the Girdle.

The Northern Triangle, or the Delta.

The fifteen constellations on the south of the ecliptic were :—

The Whale.

Orion, with the beautiful stars Rigel and Betelgeuse.

The River Endanus, or the River Orion, with the brilliant Achernar.

The Hare.

The Great Dog, with the magnificent Sirius.

The Little Dog, or the Dog which runs before, with Procyon.

The ship Argo, with its fine Alpha (Canopus) and Eta.

The Female Hydra, or the Water Snake.

The Cup, or the Urn, or the Vase.

The Raven.

The Altar, or the Perfuming Pot.

The Centaur, whose star Alpha is the nearest to the earth.

The Wolf, or the Centaur's Lance, or the Panther, or the Beast.

The Southern Crown, or the Wand of Mercury, or Uraniscus.

The Southern Fish, with Fomalhaut.

The twelve zodiacal constellations, which are of more importance than the rest, are generally named in the order in which the sun passes through them in its passage along the ecliptic, and both Latins and English have endeavoured to impress their names on the vulgar by embodying them in verses. The poet Ausonius thus catalogues them:—

"Sunt : Aries, Taurus, Gemini, Cancer, Leo, Virgo,
    Libraque, Scorpius, Arcitenens, Caper, Amphora, Pisces."

and the English effusion is as follows:—

" The Ram, the Bull, the Heavenly Twins,
    And next the Crab the Lion shines,
        The Virgin and the Scales.
    The Scorpion, Archer, and He Goat,
    The Man that holds the watering-pot,
        And Fish with glittering scales."

These twelve have hieroglyphics assigned to them, by which they are referred to in calendars and astronomical works, some of the marks being easily traced to their origin.

Thus ♈ refers to the horns of the Ram; ♉ to the head of the Bull; ♏ to the joints and tail-sting of the Scorpion; →→ is very clearly connected with an archer; ♑ is formed by the junction of the first two letters τ and ρ in $\tau\rho\acute{a}\gamma o\varsigma$, the Sea-goat, or Capricorn; ♎ for the Balance, is suggestive of its shape; ♒ refers to the water in the Watering-pot; and perhaps ♓ to the Two Fishes; ♊ for Twins may denote two sides alike; ♋ for the Crab, has something of its side-walking appearance; while ♌ for the Lion, and ♍ for the Virgin, seem to have no reference that is traceable.

These constellations contain the following stars of the first magnitude—Aldebaran, Antares, and Spica.

To these constellations admitted by the Greeks should be added the Locks of Berenice, although it is not named by Ptolemy. It was invented indeed by the astronomer Conon. The story is that Berenice was the spouse and the sister of Ptolemy Euergetes, and that she made a vow to cut off her locks and devote them to Venus if her husband returned victorious; to console the king the astronomer placed her locks among the stars. If this is a true account Arago must be mistaken in asserting that the constellation was created by Tycho Brahe in 1603. The one he did add to the former ones was that of Antinöus, by collecting into one figure some unappropriated stars near the Eagle. At about the same time J. Bayer, from the information of Vespuccius and the sailors, added twelve to the southern constellations of Ptolemy; among which may be mentioned the Peacock, the Toucan, the Phœnix, the Crane, the Fly, the Chameleon, the Bird of Paradise, the Southern Triangle, and the Indian.

Augustus Royer, in 1679, formed five new groups, among which we may name the Great Cloud, the Fleur-de-Lis, and the Southern Cross.

Hevelius, in 1690, added 16; the most important being the Giraffe, the Unicorn, the Little Lion, the Lynx, the Little Triangle.

Among these newer-named constellations none is more interesting than the Southern Cross, which is by some considered as the most brilliant of all that are known. Some

account of it, possibly from the Arabs, seems to have
reached Dante, who evidently refers to it, before it had
been named by Royer, in a celebrated passage in his
"Purgatory." Some have thought that his reference to
such stars was only accidental, and that he really referred
only to the four cardinal virtues of theology, chiefly on
account of the difficulty of knowing how he could have
heard of them; but as the Arabs had establishments along
the entire coast of Africa, there is no difficulty in under-
standing how the information might reach Italy.

Americus Vespuccius, who in his third voyage refers to
these verses of Dante, does not mention the name of the
Southern Cross. He simply says that the four stars form
a rhomboidal figure. As voyages round the Cape multiplied, however, the constellation became rapidly more
celebrated, and it is mentioned as forming a brilliant
cross by the Florentine Andrea Corsali, in 1517, and a
little later by Pigafetta, in 1520.

All these constellations have not been considered sufficient, and many subsequent additions have been made.
Thus Lacaille, in 1752, created fourteen new ones, mostly
characterized by modern names— as the Sculptor's Studio,
the Chemical Furnace, the Clock, the Compass, the Telescope, the Microscope, and others.

Lemonnier, in 1766, added the Reindeer, the Solitaire, and
the Indian Bird, and Lalande the Harvestman. Poczobut, in

1777, added one more, and P. Hell another. Finally, in the charts drawn by Bode, eight more appear, among which the Aerostat, and the Electrical and Printing Machines.

We thus arrive at a total of 108 constellations. To which we may add that the following groups are generally recognized. The Head of Medusa, near Perseus; the Pleiades, on the back, and the Hyades on the forehead of the Bull; the Club of Hercules; the Shield of Orion, sometimes called the Rake; the Three Kings; the Staff of S. James; the Sword of Orion; the Two Asses in the Crab, having between them the Star Cluster, called the Stall, or the Manger; and the Kids, near Capella, in the constellation of the Coachman.

This brings the list of the constellations to 117, which is the total number now admitted.

A curious episode with respect to these star arrangements may here be mentioned.

About the eighth century Bede and certain other theologians and astronomers wished to depose the Olympian gods. They proposed, therefore, to change the names and arrangements of the constellations; they put S. Peter in the place of the Ram; S. Andrew instead of the Bull; and so on. In more recent calendars David, Solomon, the Magi, and other New and Old Testament characters were placed in the heavens instead of the former constellations; but these changes of name were not generally adopted.

As an example of these celestial spheres we figure a
portion of one named *Cæli stellati Christiani hemispheri-
cum prius.* We here see the Great Bear replaced by the
Barque of S. Peter, the Little Bear by S. Michael, the

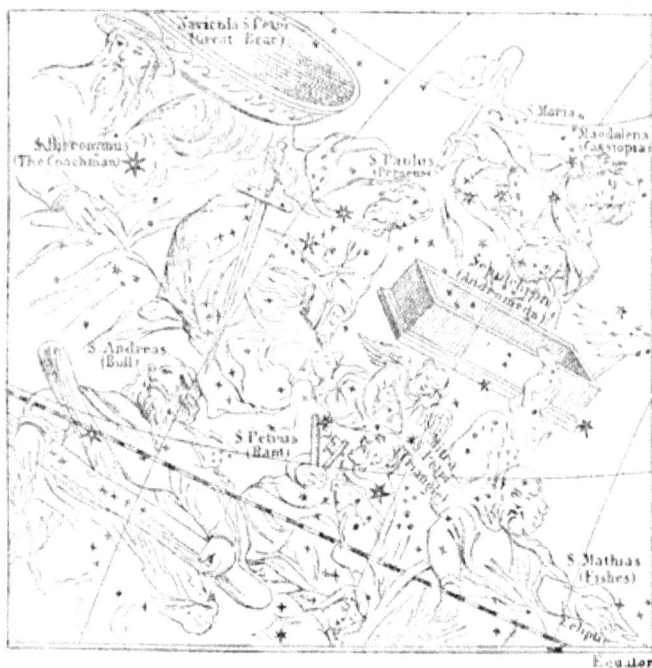

Fig. 3.

Dragon by the Innocents, the Coachman by S. Jerome,
Perseus by S. Paul, Cassiopeia by the Magdalene, Andromache
by S. Sepulchre, and the Triangle by S. Peter's mitre; while
for the zodiac were substituted the Twelve Apostles.

In the seventeenth century a proposal was made by

Weigel, a professor in the University of Jena, to form a
series of heraldic constellations, and to use for the zodiac
the arms of the twelve most illustrious families in Europe;
but these attempts at change have been in vain, the old
names are still kept.

Having now explained the origin in modern times of 69
out of the 117 constellations, there remain the 48 which
were acknowledged by the Greeks, whose origin is involved
in more obscurity.

One of the first to be noticed and named, as it is now
the most easily recognized and most widely known, is the
Great Bear, which attracts all the more attention that it is
one of those that never sets, being at a less distance from
the pole than the latter is from the horizon.

Every one knows the seven brilliant stars that form this
constellation. The four in the rectangle and the three in
a curved line at once call to mind the form of a chariot,
especially one of antique build. It is this resemblance, no
doubt, that has obtained for the constellation the name of
" the Chariot" that it bears among many people. Among
the ancient Gauls it was " Arthur's Chariot." In France it
is " David's Chariot," and in England it goes by the name
of " King Charles' Wain," and by that of the " Plough."
The latter name was in vogue, too, among the Latins (*Plau-
strum*), and the three stars were three oxen, from whence
it would appear that they extended the idea to all the seven

stars, and at last called them the *seven* oxen, *septem-triones*,
from whence the name sometimes used for the north—sep-
tentrional.   The Greeks also called it the Chariot ("Άμαξα),
and the same word seems to have stood sometimes for a
plough. It certainly has some resemblance to this instrument.

If we take the seven stars as representing the characteristic points of a chariot, the four stars of the quadrilateral
will represent the four wheels, and the three others will
represent the three horses. Above the centre of the three
horses any one with clear sight may perceive a small star
of the fifth or sixth magnitude, called the Cavalier. Each
of these several stars is indicated, as is usual with all the
constellations, by a Greek letter, the largest being denoted
by the first letter. Thus the 4 stars in the quadrilateral
are a, β, γ, δ, and the 3 tail stars ϵ, ξ, η. The Arabs give to
each star its special name, which in this case are as follows:
—Dubhé and Mérak are the stars at the back; Phegda
and Megrez those of the front; Alioth, Mizat, and Ackïar
the other three, while the little one over Mizat is Alcor.
Another name for it is Saidak, or the Tester, the being
able to see it being a mark of clear vision.

There is some little interest in the Great Bear on account
of the possibility of its being used as a kind of celestial
time-keeper, and its easy recognition makes it all the more
available. The line through a and β passes almost exactly

through the pole. Now this line revolves of course with the constellation round the pole in 24 hours; in every such interval being once vertical above the pole, and once vertical below, taking the intermediate positions to right and left between these times. The instant at which this line is vertical over the pole is not the same on any two consecutive nights, since the stars advance each day 4 minutes on the sun. On the 21st of March the superior passage takes place at 5 minutes to 11 at night; on the following night four minutes earlier, or at 9 minutes to 11. In three months the culmination takes place 6 hours earlier, or at 5 minutes to 5. In six months, i.e. on Sept. 22, it culminates at 10.55 in the morning, being vertically below the pole at the same hour in the evening. The following woodcut exhibits the positions of the Great Bear at the various hours of September 4th. It is plain from this that, knowing the day of the month, the hour of the night may be told by observing what angle the line joining a and β of this constellation makes with the vertical.

We have used the name *Great Bear*, by which the constellation is best known. It is one of the oldest names also, being derived from the Greeks, who called it Arctos megale (Ἄρκτος μεγάλη), whence the name Arctic; and singularly enough the Iroquois, when America was discovered, called it Okouari, their name for a bear. The explanation of this name is certainly not to be found in

the resemblance of the constellation to the animal. The
three stars are indeed in the tail, but the four are in the
middle of the back; and even if we take in the smaller
stars that stand in the feet and head, no ingenuity can

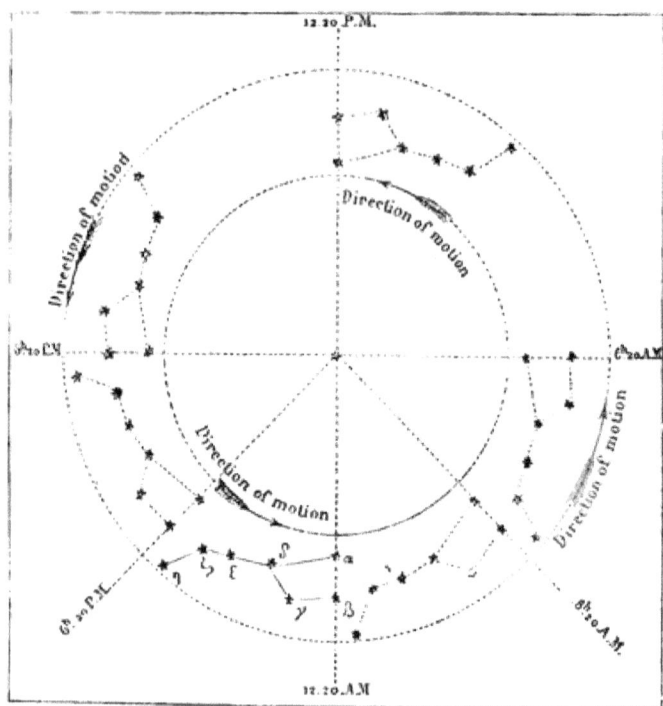

FIG. 4.

make it in this or any other way resemble a bear. It
would appear, as Aristotle observes, that the name is derived
from the fact, that of all known animals the bear was
thought to be the only one that dared to venture into the

frozen regions of the north and tempt the solitude and
cold.

Other origins of the name, and other names, have been
suggested, of which we may mention a few. For example,
" Ursa" is said to be derived from *versus*, because the con-

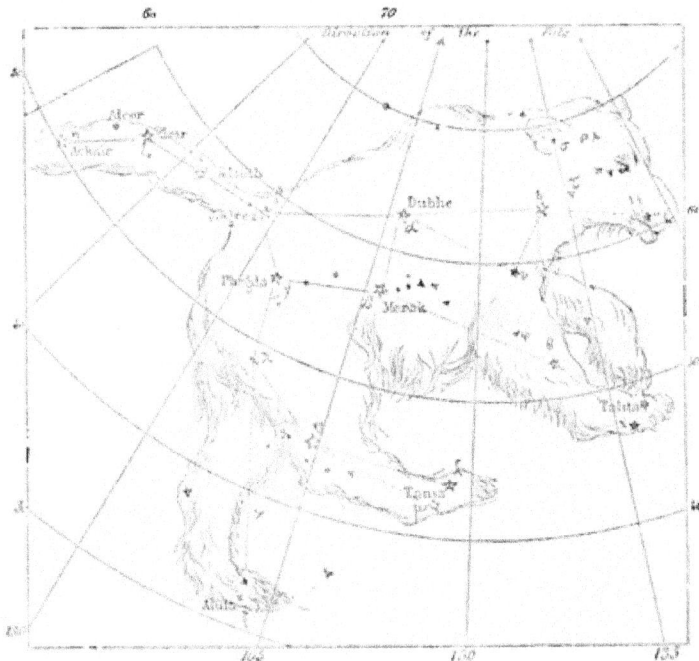

Fig. 5.

stellation is seen to *turn* about the pole. It has been called
the Screw ("Ελικη), or Helix, which has plainly reference
to its turning. Another name is Callisto, in reference to
its beauty; and lastly, among the Arabs the Great and
Little Bears were known as the Great and Little Coffins

in reference to their slow and solemn motion. These names referred to the four stars of each constellation, the other three being the mourners following the bearers. The Christian Arabs made it into the grave of Lazarus and the three weepers, Mary, Martha, and their maid.

Next as to the Little Bear. This constellation has evidently received its name from the similarity of its form to that of the Great Bear. In fact, it is composed of seven stars arranged in the same way, only in an inverse order. If we follow the line from β to a of the Great Bear to a distance of five times as great as that between these stars we reach the brightest star of the Little Bear, called the Pole Star. All the names of the one constellation have been applied to the other, only at a later date.

The new constellations were added one by one to the celestial sphere by the Greeks before they arranged certain of them as parts of the zodiac. The successive introduction of the constellations is proved completely by a long passage of Strabo, which has been often misunderstood. "It is wrong," he says, "to accuse Homer of ignorance because he speaks only of one of the two Celestial Bears. The second was probably not formed at that time. The Phenicians were the first to form them and to use them for navigation. They came later to the Greeks."

All the commentators on Homer, Hygin and Diogenes Laertes, attribute to Thales the introduction of this constel-

THE CONSTELLATIONS FROM THE SEA-SHORE.

The Swan—The Lyre—Hercules—The Crown—The Herdsman—The
Eagle—The Serpent—The Balance—The Scorpion—Sagittarius.

lation. Pseudo-Eratosthenes called the Little Bear Φοινίκη, to indicate that it was a guide to the Phenicians. A century later, about the seventeenth Olympiad, Cleostratus of Tenedos enriched the sphere with the Archer (Τοξότης, Sagittarius) and the Ram (Κριός, Aries), and about the same time the zodiac was introduced into the Grecian sphere.

With regard to the Little Bear there is another passage of Strabo which it will be interesting to quote. He says— "The position of the people under the parallel of Cinnamomophore, i.e. 3,000 stadia south of Meroe and 8,800 stadia north of the equator, represents about the middle of the interval between the equator and the tropic, which passes by Syene, which is 5,000 stadia north of Meroe. These same people are the first for whom the Little Bear is comprised entirely in the Arctic circle and remains always visible; the most southern star of the constellation, the brilliant one that ends the tail being placed on the circumference of the Arctic circle, so as just to touch the horizon." The remarkable thing in this passage is that it refers to an epoch anterior to Strabo, when the star a of the Little Bear, which now appears almost immovable, owing to its extreme proximity to the pole, was then more to the south than the other stars of the constellation, and moved in the Arctic circle so as to touch the horizon of places of certain latitudes, and to set for latitudes nearer the equator.

F

In those days it was not the *Pole* Star—if that word has any relation to πολέω, I turn—for the heavens did not turn about it then as they do now.

The Grecian geographer speaks in this passage of a period when the most brilliant star in the neighbourhood of the pole was *a* of the Dragon. This was more than three thousand years ago. At that time the Little Bear was nearer to the pole than what we now call the Polar Star, for this latter was "the most southern star in the constellation." If we could alight upon documents dating back fourteen thousand years, we should find the star Vega (*a* Lyra) referred to as occupying the pole of the world, although it now is at a distance of 51 degrees from it, the whole cycle of changes occupying a period of about twenty-six thousand years.

Before leaving these two constellations we may notice the origin of the names according to Plutarch. He would have it that the names are derived from the use that they were put to in navigation. He says that the Phenicians called that constellation that guided them in their route the *Dubic*, or *Doubc*, that is, the speaking constellation, and that this same word happens to mean also in that language a bear; and so the name was confounded. Certainly there is still a word *dubbeh* in Arabic having this signification.

Next as to the Herdsman. The name of its characteristic star and of itself, Arcturus (Ἄρκτος, bear; Οὖρος, guardian), is explained without difficulty by its position near the Bears. There are six small stars of the third magnitude in the constellation round its chief one—three of its stars forming an equilateral triangle. Arcturus is in the continuation of the curved line through the three tail stars of the Great Bear. The constellation has also been called Atlas, from its nearness to the pole—as if it held up the heavens, as the fable goes.

Beyond this triangle, in the direction of the line continued straight from the Great Bear, is the Northern Crown, whose form immediately suggests its name. Among the stars that compose it one, of the second magnitude, is called the Pearl of the Crown. It was in this point of the heavens that a temporary star appeared in May, 1866, and disappeared again in the course of a few weeks.

Among the circumpolar constellations we must now speak of Cassiopeia, or the Chair—or Throne—which is situated on the opposite side of the Pole from the Great Bear; and which is easily found by joining its star δ to the Pole and continuing it. The Chair is composed principally of five stars, of the third magnitude, arranged in the form of an M. A smaller star of the fourth magnitude completes the square formed by the three β, a, and γ. The figure thus formed has a fair resemblance to a chair or throne, δ and ϵ forming the back; and hence the justification for its popular name. The other name Cassiopeia has its connection and meaning unknown.

We may suitably remark in this place, with Arago, that no precise drawing of the ancient constellations has come down to us. We only know their forms by written descriptions, and these often very short and meagre. A verbal description can never take the place of a drawing, especially if it is a complex figure, so that there is a certain amount of doubt as to the true form, position, and arrangement of the figures of men, beasts, and inanimate objects which composed the star-groups of the Grecian astronomers—so that unexpected difficulties attend the attempt to reproduce them on our modern spheres. Add to this that alterations have been avowedly introduced by the ancient astronomers themselves, among others by Ptolemy, especially in those given by Hipparchus. Ptolemy says he determined to make these changes because it was necessary to give a better proportion to the figures, and to adapt them better to the real positions of the stars. Thus in the constellation of the Virgin, as drawn by Hipparchus, certain stars corresponded to the shoulders ; but Ptolemy placed them in the sides, so as to make the figure a more beautiful one. The result is that modern designers give scope to their imagination rather than consult the descriptions of the Greeks. *Cassiopeia, Cepheus, Andromeda*, and *Perseus* holding in his hand the *Head of Medusa*, appear to have been established at the same epoch, no doubt subsequently to the Great Bear. They form one family, placed together in one part of the heavens, and associated in one drama; the ardent Perseus delivering the unfortunate Andromeda, daughter of Cepheus and Cassiopeia. We can never be sure, however, whether the constellations suggested the fable, or the fable the constellations: the former may only mean that Perseus, rising before Andromeda, seems to deliver it from the Night and from the constellation of the Whale. The Head of Medusa, a celebrated woman, that Perseus cut off and holds in his hand, is said by Volney to be only the head of the constellation Virgo, which passes beneath the horizon precisely as the Perseus rises, and the serpents which surround it are Ophiucus and the polar Dragon, which then occupies the zenith.

Either way, we have no account of the origin of the *names*, and it is possible that we may have to seek it, if ever we find it, from other sources—for it would appear that similar names were used for the same constellations by the Indians. This seems inevitably proved by what is related by Wilford (*Asiatic Researches*, III.) of his conversation with his pundit, an astronomer, on the names of the Indian constellations. "Asking him," he says, "to show me in the heavens the constellation of Antarmada, he immediately pointed to Andromeda, though I had not given him any information about it beforehand. He afterwards brought me a very rare and curious work in Sanscrit, which contained a chapter devoted to *Upanacchatras*, or extra-zodiacal constellations, with drawings of *Capuja* (Cepheus), and of *Casyapi* (Cassiopeia) seated and holding a lotus flower in her hand, of *Antarmada* charmed with the fish beside her, and last of *Parasica* (Perseus) who, according to the explanation of the book, held the head of a monster which he had slain in combat ; blood was dropping from it, and for hair it had snakes." As the stars composing a constellation have often very little connection with the figure they are supposed to form, when we find the same set of stars called by the same name by two different nations, as was the case, for instance, in some of the Indian names of constellations among the Americans, it is a proof that one of the nations copied it from the other, or that both have copied from a common source. So in the case before us, we cannot think these similar names have arisen independently, but must conclude that the Grecian was borrowed from the Indian.

Another well-known constellation in this neighbourhood, forming an isosceles triangle with Arcturus and the Pole Star, is the Lyre. Lucian of Samosatus says that the Greeks gave this name to the constellation to do honour to the Lyre of Orpheus. Another possible explanation is this. The word for lyre in Greek ($\chi \acute{\epsilon} \lambda \upsilon \varsigma$) and in Latin (*testudo*) means also a tortoise. Now at the time when this name was imposed the chief star in the Lyre may have been very near to the pole of the heavens and therefore have had a very slow motion, and hence it might have been named the tortoise, and this in Greek would easily be interpreted into lyre instead. Indeed this double meaning of the word seems certainly to have given rise to the fable of Mercury having

constructed a lyre out of the back of a tortoise. Circling round
the pole of the ecliptic, and formed by a sinuous line of
stars passing round from the Great Bear to the Lyre, is the
Dragon, which owes its name to its form. Its importance is
derived from its relation to the ecliptic, the pole of which is
determined by reference to the stars of the first coil of the
body. The centre of the zodiacal circle is a very important
point, that circle being traced on the most ancient spheres, and
probably being noticed even before the pole of the heavens.

Closely associated with the Dragon both in mythology and
in the celestial sphere is Hercules. He is always drawn
kneeling; in fact, the constellation is rather a man in a
kneeling posture than any particular man. The poets
called it Engonasis with reference to this, which is too
melancholy or lowly a position than would agree well with
the valiant hero of mythology. There is a story related by
Æschylus about the stones in the Champ des Cailloux, between
Marseilles and the embouchure of the Rhône, to the effect that
Hercules, being amongst the Ligurians, found it necessary to
fight with them; but he had no more missiles to throw; when
Jupiter, touched by the danger of his son, sent a rain of round
stones, with which Hercules repulsed his enemies. The Engonasis is thus considered by some to represent him bending
down to pick up the stones. Posidonius remarks that it was
a pity Jupiter did not rain the stones on the Ligurians at
once, without giving Hercules the trouble to pick them up.

Ophiucus, which comes close by, simply means the man that holds the serpent (ὀφι-οῦχος).

It is obviously impossible to know the origins of all the names, as those we now use are only the surviving ones of several that from time to time have been applied to the various constellations according to their temporary association with the local legends. The prominent ones are favoured with quite a crowd of names. We need only cite a few. Hercules, for instance, has been called Ὀκάλζων Κορυνήτης, Engonasis, Ingeniculus, Nessus, Thamyris, Desanes, Maceris, Almannus, Al-chete, &c. The Swan has the names of Κύκνος, Ἴκτιν, Ὄρνις, Olar, Helenæ genitor, Ales Jovis, Ledæus, Milvus, Gallina, The Cross, while the Coachman has been Ἱππιλατης, Ἐλαστίππος, Αἰρωηλατης, Ἡνιοχος, Auriga, Acator, Hemochus, Erichthonus, Mamsek, Alánat, Athaiot, Alatod, &c. With respect to the Coachman, in some old maps he is drawn with a whip in his left hand turned towards the chariot, and is called the charioteer. No doubt its proximity to the former constellation has acquired for it its name. The last we need mention, as of any celebrity, is that of Orion, which is situated on the equator, which runs exactly through its midst. Regel forms its left foot, and the Hare serves for a footstool to the right foot of the hero. Three magnificent stars in the centre of the quadrilateral, which lie in one straight line are called the Rake, or the Three Kings, or the Staff of Jacob, or the Belt.

These names have an obvious origin; but the meaning of Orion itself is more doubtful. In the Grecian sphere it is

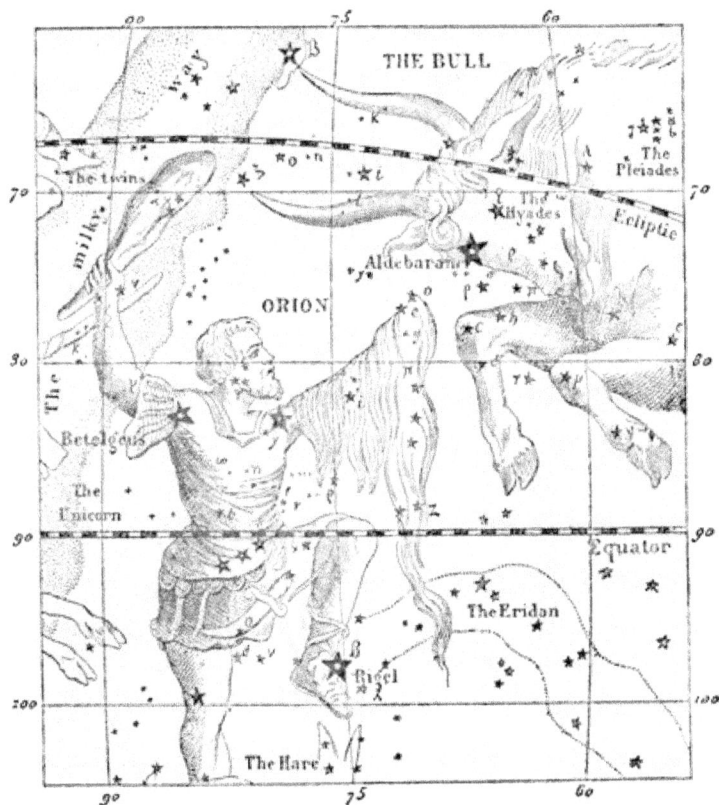

FIG. 6.

written Ὠρίων, which also means a kind of bird. The allied word ὧρος has very numerous meanings, the only one of which that could be conjectured to be connected with the constellations is a "guardian." The word ἴριον, on the

contrary, the diminutive of ὥρος, means a limit, and has
been assigned to Jupiter; and in this case may have
reference to the constellation being situated on the confines
of the two hemispheres.   In mythology Orion was an
intrepid hunter of enormous size.   He was the same
personage as Orus, Arion, the Minotaur, and Nimrod, and
afterwards became Saturn.   Orion is called *Tsan* in Chinese,
which signifies three, and corresponds to the three kings.

The Asiatics used not to trace the images of their
constellations, but simply joined the component stars by
straight lines, and placed at the side the hieroglyphic
characters that represented the object they wished to name.
Thus joining by five lines the principal stars in Orion, they
placed at the side the hieroglyphics representing a man
and a sword, from whence the Greeks derived the figure they
afterwards drew of a giant armed with a sword.

We must include in this series that brightest of all stars,
Sirius.   It forms part of the constellation of the Great Dog,
and lies to the south of Orion near the extreme limit of
our vision into the Southern hemisphere in our latitudes.
This star seems to have been intimately connected with
Egypt, and to have derived its name—as well as the name
of the otherwise unimportant constellation it forms part of—
from that country, and in this way :—

The overflowing of the Nile was always preceded by an
Etesian wind, which, blowing from north to south about the

time of the passage of the sun beneath the stars of the Crab, drove the mists to the south, and accumulated them over the country whence the Nile takes its source, causing abundant rains, and hence the flood. The greatest importance attached to the foretelling the time of this event, so that people might be ready with their provisions and their places of security. The moon was no use for this purpose, but the stars were, for the inundation commenced when the sun was in the stars of the Lion. At this time the stars of the Crab just appeared in the morning, but with them, at some distance from the ecliptic, the bright star Sirius also rose. The morning rising of this star was a sure precursor of the inundation. It seemed to them to be the warning star, by whose first appearance they were to be ready to move to safer spots, and thus acted for each family the part of a faithful dog. Whence they gave it the name of the Dog, or Monitor, in Egyptian *Anubis*, in Phenician *Hannobeach*, and it is still the Dog-Star—*Caniculus*, and its rising commences our *dog-days*. The intimate connection between the rising of this star and the rising of the Nile led people to call it also the Nile star, or simply the Nile; in Egyptian and Hebrew, *Sihor ;* in Greek, Σοθίς; in Latin, *Sirius.*

In the same way the Egyptians and others characterised the different days of the year by the stars which first appeared in the evening—as we shall see more particularly with reference to the Pleiades—and in this way certain stars came to be associated in their calendar with variations of temperature and operations of agriculture. They soon took for the cause what was originally but the sign, and thus they came to talk of moist stars, whose rising brought rain, and arid stars, which brought drought. Some made certain plants to grow, and others had influence over animals.

In the case of Egypt, no other so great event could occur as that which the Dog-Star foretold, and its appearance was consequently made the commencement of the year. Instead, therefore, of painting it as a simple star, in which case it would be indistinguishable from others, they gave it shape according to its function and name. When they wished to signify that it opened the year, it was represented as a porter bearing keys, or else they gave it two heads, one of an old man, to represent the passing year, the other of a younger, to denote the succeeding year. When they would represent it as giving warning of the inundation they painted it as a dog. To illustrate what they were to do when it appeared, Anubis had in his arms a stew-pot, wings to his feet, a large feather under his arm, and two reptiles behind him, a tortoise and a duck.

There is also in the celestial sphere a constellation called the Little Dog and Procyon; the latter name has an obvious meaning, as appearing *before* the Dog-Star.

We cannot follow any farther the various constellations of the northern sphere, nor of the southern. The zodiacal

constellations we must reserve for the present, while we
conclude by referring to some of the changes in form and
position that some of the above-mentioned have undergone
in the course of their various representations.

These changes are sometimes very curious, as, for example,
in a coloured chart, printed at Paris in 1650, we have the
Charioteer drawn in the costume of Adam, with his knees on
the Milky Way, and turning his back to the public ; the she-
goat appears to be climbing over his neck, and two little
she-goats seem to be running towards their mother.
Cassiopeia is more like King Solomon than a woman.
Compare this with the *Phenomena of Aratus*, published
1559, where Cassiopeia is represented sitting on an oak
chair with a ducal back, holding the holy palm in her left
hand, while the Coachman, " Erichthon," is in the costume of
a minion of Henry the Third of France. Now compare the
Cassiopeia of the Greeks with that drawn in the sixteenth
and seventeenth centuries, or the Coachman of the same
periods, and we can easily see the fancies of the painters
have been one of the most fertile sources of change. They
seem, too, to have had the fancy in the middle ages to draw
them all hideous and turning their backs. Compare, for
instance, the two pictures of Andromeda and Hercules, as
given below, where those on one side are as heavy and gross
as the others are artistic and pretty. Unfortunately for the
truth of Andromeda's beauty, as depicted in these designs,

she was supposed to be a negress, being the daughter of the Ethiopians, Cepheus and Cassiopeia. Not one of the

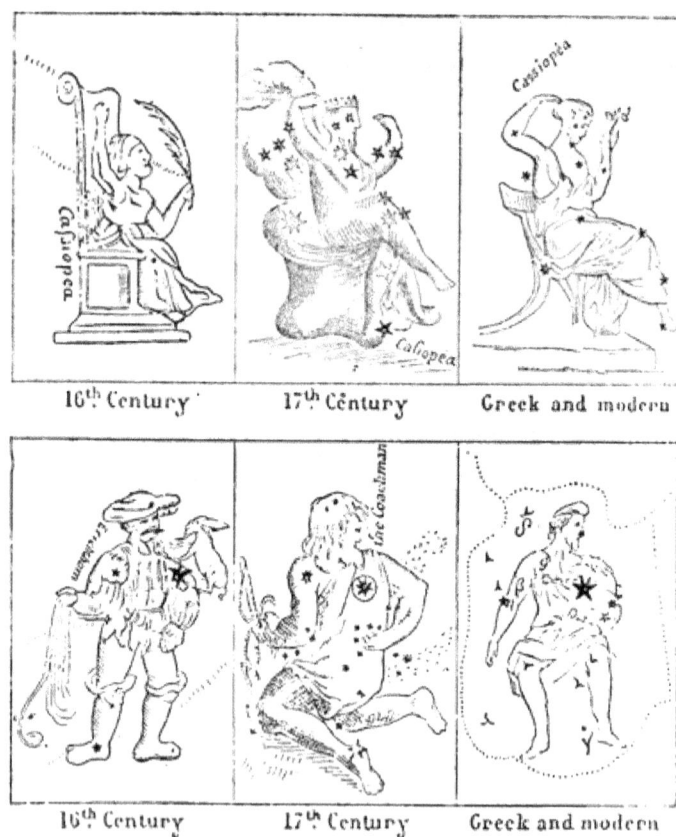

Fig. 7.

drawings indicates this; indeed they all take after their local beauties.

In Flamsteed's chart, as drawn above, the Coachman is

a female; and instead of the she-goat being on the back, she

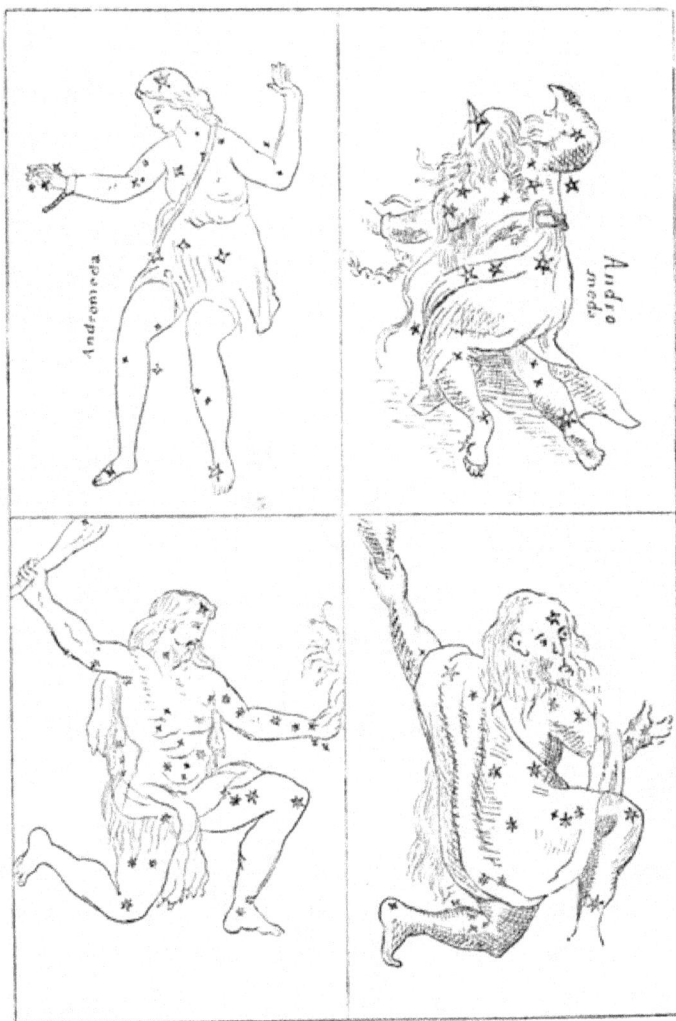

Fig. 8.

holds it in her arms. No one, indeed, from any of the

figures of this constellation would ever dream it was intended to represent a coachman.

One more fundamental cause of changes has been the confusion of names derived by one nation from another, these having sometimes followed their signification, but at others being translated phonetically. Thus the Latins, in deriving names from the Greek Ἄρκτος, have partly translated it by Ursa, and partly have copied it in the form Arcticus. So also with reference to the three stars in the head of the Bull, called by the Greeks Hyades. The Romans thought it was derived from ὕες, sows, so they called them *suculæ*, or little sows; whereas the original name was derived from ὕειν, to rain, and signified stars whose appearance indicated the approach of the rainy season.

More curious still is the transformation of the Pearl of the Northern Crown (Margarita Coronæ) in a saint— S. Marguerite.

The names may have had many origins whose signification is lost, owing to their being misunderstood. Thus figurative language may have been interpreted as real, as when a conjunction is called a marriage; a disappearance, death; and a reappearance, a resurrection; and then stories must be invented to fit these words; or the stars that have in one country given notice of certain events lose the meaning of their names when these are used elsewhere; as when a boat painted near the stars that accompany an inundation, becomes the ship Argo; or when, to represent the wind, the bird's wing is drawn; or those stars that mark a season are associated with the bird of passage, the insect or the animal that appears at that time : such as these would soon lose their original signification.

The celestial sphere, therefore, as we now possess it, is not simply a collection of unmeaning names, associated with a group of stars in no way connected with them, which have been imposed at various epochs by capricious imagination, but in most instances, if not in all, they embody a history, which, if we could trace it, would probably lead us to astronomical facts, indicating the where and the when of their first introduction; and the story of their changes, so far as we can trace it, gives us some clue to the mental characteristics or astronomical progress of the people who introduced the alterations.

We shall find, indeed, in a subsequent chapter, that many of our conclusions as to the birth and growth of astronomy are derived from considerations connected with the various constellations, more especially those of the zodiac.

With regard to the date when and the country where the constellations of the sphere were invented, we will here give what evidence we possess, independent of the origin of the zodiac.

In the first place it seems capable of certain proof that they were not invented by the Greeks, from whom we have

G

received them, but adopted from an older source, and it is
possible to give limits to the date of introduction among
them.

Newton, who attributes its introduction to Musæus, a contemporary of Chiron, remarks, that it must have been settled
after the expedition of the Argonauts, and *before* the destruction of Troy; because the Greeks gave to the constellation
names that were derived from their history and fables, and
devoted several to celebrate the memory of the famous
adventurers known as the Argonauts, and they would
certainly have dedicated some to the heroes of Troy, if the
siege of that place had happened at the time. We remark
that at this time astronomy was in too infant a state in
Greece for them to have fixed with so much accuracy the
position of the stars, and that we have in this a proof
they must have borrowed their knowledge from older
cultivators of the science.

The various statements we meet with about the invention
of the sphere may be equally well interpreted of its introduction only into Greece. Such, for instance, as that Eudoxus
first constructed it in the thirteenth or fourteenth century
B.C., or that by Clement of Alexandria, that Chiron was
the originator.

The oldest direct account of the names of the constellations
and their component stars is that of Hesiod, who cites by
name in his *Works and Days* the Pleiades, Arcturus, Orion,

and Sirius. He lived, according to Herodotus, about 884
years before Christ.

The knowledge of all the constellations did not reach the
Greeks at the same time, as we have seen from the omission
by Homer of any mention of the Little Bear, when if he
had known it, he could hardly have failed to speak of it.
For in his description of the shield of Achilles, he mentions
the Pleiades, the Hyades, Orion and the Bear, " which alone
does not bathe in the Ocean." He could never have said
this last if he had known of the Dragon and Little Bear.

We may then safely conclude that the Greeks received the
idea of the constellations from some older source, probably
the Chaldeans. They received it doubtless as a sphere, with
figured, but nameless constellations ; and the Greeks by slight
changes adapted them to represent the various real or
imaginary heroes of their history. It would be a gracious
task, for their countrymen would glory in having their great
men established in the heavens. When they saw a ship
represented, what more suitable than to name it the ship
Argo? The Swan must be Jupiter transformed, the Lyre
is that of Orpheus, the Eagle is that which carried away
Ganymede, and so on.

This would be no more than what other nations have
done, as, for example, the Chinese, who made greater changes
still, unless we consider theirs to have had an entirely
independent origin.

G 2

That the celestial sphere was a conception known to others than the Greeks is easily proved. The Arabians, for instance, certainly did not borrow it from them; yet they have the

Fig. 9.

same things represented. Above is a figure of a portion of an Arabian sphere drawn in the eleventh century, where we get represented plainly enough the Great and Little Bears, the Dragon, Cassiopeia, Andromeda, Perseus, with the Triple Head of Medusa ; the Triangle, one of the Fishes, Auriga, the Ram, the Bull obscurely, and the Twins.

There is also the famous so-called zodiac of Denderah, brought from Egypt to Paris. This in reality contains more constellations than those of the zodiac. Most of the northern ones can be traced, with certain modifications. Its construction is supposed to belong to the eighth century B.C. Most conspicuous on it is the Lion, in a kind of barque, recalling the shape of the Hydra. Below it is the calf Isis, with Sirius, or the Dog-Star, on the forehead ; above it is the Crab, to the right the Twins, over these a long instrument, the Plough, and above that a small animal, the Little Bear, and so we may go on:—all the zodiacal constellations, especially the Balance, the Scorpion, and the Fishes being very clear. This sphere is indeed of later date than that supposed for the Grecian, but it certainly appears to be independent. The remains we possess of older spheres are more particularly connected with the zodiac, and will be discussed hereafter.

From what people the Greeks received the celestial sphere, is a question on which more than one opinion has been formed. One is that it was originated in the tropical latitudes of Egypt. The other, that it came from the Chaldeans, and a third that it came from more temperate latitudes further to the east. The arguments for the last of these are as follow :

There is an empty space of about 90°, formed by the last
constellations of the sphere, towards the south pole, that
is by the Centaur, the Altar, the Archer, the Southern Fish,
the Whale, and the Ship. Now in a systematic plan, if the
author were situated near the equator there would be no
vacant space left in this way, for in this case the southern
stars, attracting as much attention as the northern, would be
inevitably inserted in the system of constellations which
would be extended to the horizon on all sides. But a
country of sufficiently high latitude to be unable to see at
any time the stars about the southern pole must be north
both of Egypt and Chaldea.

This empty space remained unfilled until the discovery of
the Cape of Good Hope, except that the star Canopus was
included in the constellation Argo, and the river Eridan had
an arbitrary extension given to it, instead of terminating
in latitude 40°.

Another less cogent argument is derived from the interpretation of the fable of the Phœnix. This is supposed to
represent the course of the sun, which commences its growth
at the time of its death. A similar fable is found among the
Swedes. Now a tropical nation would find the difference of
days too little to lead it to invent such a fable to represent it.
It must needs have arisen where the days of winter were
very much shorter than those of summer.

The Book of Zoroaster, in which some of the earliest notices

PLATE III.—CHALDEAN ASTRONOMERS.

of astronomy are recorded, states that the length of a summer
day is twice as long as that of winter. This fixes the latitude in which that book must have been composed, and makes
it 49°. Whence it follows, that to such a place must we look
for the origin of these spheres, and not to Egypt or Chaldea.

Diodorus Siculus speaks of a nation in that part of the
world, whom he calls Hyperboreans, who had a tradition that
their country is the nearest to the moon, on which they
discovered mountains like those on the earth, and that Apollo
comes there once every nineteen years. This period being that
of the metonic cycle of the moon, shows that if this could have
really been discovered by them, they must have had a long
acquaintance with astronomy.

The Babylonian tablets lead us to the belief that astronomy,
and with it the sphere, and the zodiac were introduced by a
nation coming from the East, from the mountains of Elam,
called the Accadians, before 3000 B.C., and these may have
been the nation to whom the whole is due.

On the other hand, the arguments for the Egyptians, or
Chaldeans being the originators depend solely on the tradition handed down by many, that one or other are the oldest
people in the world, with the oldest civilization, and they
have long cultivated astronomy. More precise information,
however, seems to render these traditions, to say the least,
doubtful, and certainly incapable of overthrowing the arguments adduced above.

CHAPTER IV.

THE zodiac, as already stated, is the course in the heavens apparently pursued by the sun in his annual journey through the stars. Let us consider for a moment, however, the series of observations and reflections that must have been necessary to trace this zone as representing such a course.

First, the diurnal motion of the whole heavens from east to west must have been noticed during the night, and the fact that certain stars never set, but turn in a circle round a fixed point. What becomes then, the next question would be, of those stars that do descend beneath the horizon, since they rise in the same relative positions as those in which they set. They could not be thought to be destroyed, but must complete the part of the circle that is invisible *beneath the earth.* The possibility of any stars finding a path beneath the earth must have led inevitably to the conception of the earth as a body suspended in the centre with nothing to support it. But leaving this alone, it would also be concluded that the sun went with the stars, and was in a certain position among them, even when both they and it were invisible. The next observations necessary would be that the zodiacal constellations visible during the nights of winter were not the same as those seen in summer, that such and such a group of stars passed the meridian at midnight at a certain time, and that six months afterwards the group exactly opposite in the heavens passed at the same hour. Now since at midnight the sun will be exactly opposite the meridian, if it continues uniformly on its course, it will be among that group of stars that is opposite the group that culminates at midnight, and so the sign of the zodiac the sun occupies would be determined.

This method would be checked by comparisons made in the morning and evening with the constellations visible nearest to the sun at its rising and setting.

The difficulty and indirectness of these observations would make it probable that originally the zodiac would be determined rather by the path of the moon, which follows nearly the same path as the sun, and which could be observed at the same time as, and actually associated with, the constellations. Now the moon is found each night so far to the east of its position on the previous night that it accomplishes the whole circumference in twenty-seven days eight hours. The two nearest whole number of days have generally been reckoned,.

some taking twenty-eight, and others twenty-seven. The zodiac, or, as the Chinese called it, the Yellow Way, was thus divided into twenty-eight parts, which were called *Nakshatras* (mansions, or hotels), because the moon remains in each of them for a period of twenty-four hours. These mansions were named after the brightest stars in each, though sometimes they went a long way off to fix upon a characteristic star, as in the sixteenth Indian constellation, *Vichaca*, which was named after the Northern Crown, in latitude 40°. This arose from the brightness of the moon extinguishing the light of those that lie nearest to it.

This method of dividing the zodiac was very widely spread, and was common to almost all ancient nations. The Chinese have twenty-eight constellations, but the word *siou* does not mean a group of stars, but simply a mansion or hotel. In the Coptic and ancient Egyptian the word for constellation has the same meaning. They also had twenty-eight, and the same number is found among the Arabians, Persians, and Indians. Among the Chaldeans, or Accadians, we find no sign of the number twenty-eight. The ecliptic or "Yoke of the Sky," with them, as we see in the newly-discovered tablets, was divided into twelve divisions as now, and the only connection that can be imagined between this and the twenty-eight is the opinion of M. Biot, who thinks that the Chinese had originally only twenty-four mansions, four more being added by Chenkung (B.C. 1100), and that they corresponded with the twenty-four stars, twelve to the north and twelve to the south, that marked the twelve signs of the zodiac among the Chaldeans. But under this supposition the twenty-eight has no reference to the moon, whereas we have every reason to believe that it has.

The Siamese only reckoned twenty-seven, and occasionally inserted an extra one, called *Abigitten*, or intercalary moon. They made use, moreover, of the constellations to tell the hour of the night by their position in the heavens, and their method of doing this appears to have involved their having twenty-eight constellations. The names of the twenty-eight divisions among the Arabs were derived from parts of the larger constellations that made the twelve signs, the first being the horns, and the second the belly, of the Ram.

The twenty-eight divisions among the Persians, of which we may notice that the second was formed by the Pleiades, and called *Perris*, soon gave way to the twelve, the names of which, recorded in the works of Zoroaster, and therefore not less ancient than he, were not quite the same as those now used. They were the Lamb, the Bull, the Twins, the Crab, the Lion, the Ear of Corn, the Balance, the Scorpion, the Bow, the Sea-Goat, the Watering-pot, and the Fishes.

Nor were the Chinese continually bound to the number twenty-eight. They, too, had a zodiac for the sun as well as the moon, as may be seen on some very curious pieces of money, of which those figured below are specimens.

On some of these the various constellations of the Northern hemisphere are engraved, especially the Great Bear—under innumerable disguises—and on others the twelve signs of the zodiac. These are very different, however, from the Grecian set—they are the Mouse, the Bull, the Tiger, the Hare, the Dragon, the Serpent, the Horse, the Ram, the Monkey, the Cock, the Dog, and the Pig. The Japanese series were the

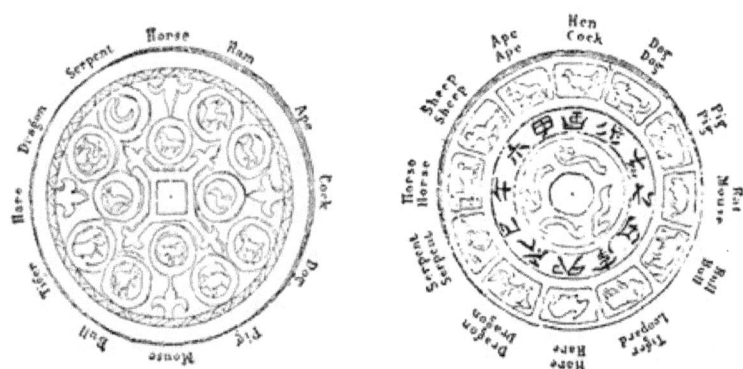

FIG. 10.

same. The Mongolians had a series of zodiacal coins struck in the reign of Jehanjir Shah (1014). He had pieces of gold stamped, representing the sun in the constellation of the Lion; and some years afterwards other coins were made, with one side having the impress of the particular sign in which the sun happened to be when the coin was struck. In this way a series is preserved having all the twelve signs. Tavernier tells the story that one of the wives of the Sultan,

wishing to immortalise herself, asked Jehanjir to be allowed
to reign for four-and-twenty hours, and took the opportunity
to have a large quantity of new gold and silver zodiacal coins
struck and distributed among the people.

The twenty-eight divisions are less known now, simply
from the fact that the Greeks did not adopt them; but they
were much used by the early Asian peoples, who distinguished
them, like the twelve, by a series of animals, and they are
still used by the Arabs.

So far for the nature of the zodiac, as used in various
countries, and as adopted from more ancient sources by the
Greeks and handed on to us. It is very remarkable that
the arrangement of it, and its relation to the pole of the
equator, carries with it some indication of the age in which
it must have been invented, as we now proceed to show.

We may remark, in the first place, that from very early
times the centre of the zodiacal circle has been marked in
the celestial sphere, though there is no remarkable star near
the spot; and the centre of the equatorial circle, or pole, has
been even less noticed, though much more obvious. We
cannot perhaps conclude that the instability of the pole was
known, but that the necessity for drawing the zodiac led to
attention being paid to its centre. Both the Persians and
the Chinese noted in addition four bright stars, which they
said watched over the rest, *Taschter* over the east, *Sateris*
over the west, *Venaud* over the south, and *Hastorang* over

the north. Now we must understand these points to refer to
the sun, the east being the spring equinox, the west the
autumnal, and the north and south the summer and winter
solstices.   There are no stars of any brilliancy that we could
now suppose referred to in these positions; but if we turn
the zodiac through 60° we shall find Aldebaran, the Antares,
Regulus, and Fomalhaut, four stars of the first magnitude,
pretty nearly in the right places.   Does the zodiac then turn
in this way ?   The answer is, It does.

The effect of the attraction of the sun and moon upon the
equatorial protuberance of the earth is to draw it round from
west to east by a very slow motion, and make the ecliptic
cross the equator each year about one minute of arc to the
east of where it crossed it the year before.   So, then, the
sidereal year, or interval between the times at which the sun
is in a certain position amongst the stars, is longer than the
solar year, or interval between the times at which the sun
crosses the equator at the vernal equinox.   Now the sun's
position in the zodiac refers to the former, his appearance at
the equinox to the latter kind of year.   Each solar year then—
and these are the years we usually reckon by—the equinox is
at a point fifty seconds of arc to the east on the zodiac, an
effect which is known by the name of the precession of the
equinoxes.

Now it is plain that if it keeps moving continuously to
the east it will at last come round to the same point again,

and the whole period of its revolution can easily be calculated from the distance it moves each year. The result of such a calculation shows that the whole revolution is completed in 25,870 years, after which time all will be again as it is now in this respect.

If we draw a figure of the zodiac, as below, and know

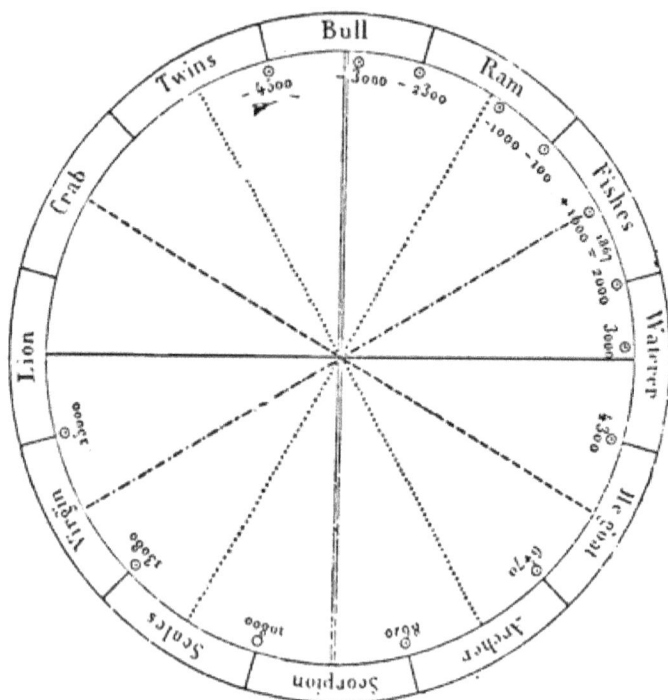

Fig. 11.

that at this time the vernal equinox takes place when the sun is in the Fishes, then, the constellation of the Ram being

to the west of this, the date at which the equinox was there
must be before our present date, while at some time in the
future it will be in Aquarius.

Now if in any old description we find that the equinox is
referred to as being in the Ram or in the Bull, it tells us at
once how long ago such a description was a true one, and,
therefore, when it was written. This is the way in which
the Zodiac carries with it an intimation of its date. Thus
in the example lately referred to of the Persians and their
four stars, it must have been about 5,000 years ago, according
to the above calculation, that these were in the positions
assigned, which is therefore the date of this part of Persian
astronomy, if we have rightly conjectured the stars
referred to.

We have already said that the signs of the zodiac are not
now the same as the zodiacal constellations, and this is now
easily understood. It is not worth while to say that the
sun enters such and such a part of the Fishes at the equinox,
and changes every year. So the part of the heavens it *does*
then enter—be it Fishes, or Aquarius, or the Ram—is called
by the same name—and is called a *sign*; the name chosen is
the Ram or Aries, which coincided with the constellation of
that name when the matter was arranged. There is another
equally important and instructive result of this precession of
the equinox. For the earth's axis is always perpendicular to
the plane of the equator, and if the latter moves, the former

must too, and change its position with respect to the axis of
the ecliptic, which remains immovable.   And the ends of
these axes, or the points they occupy among the stars, called

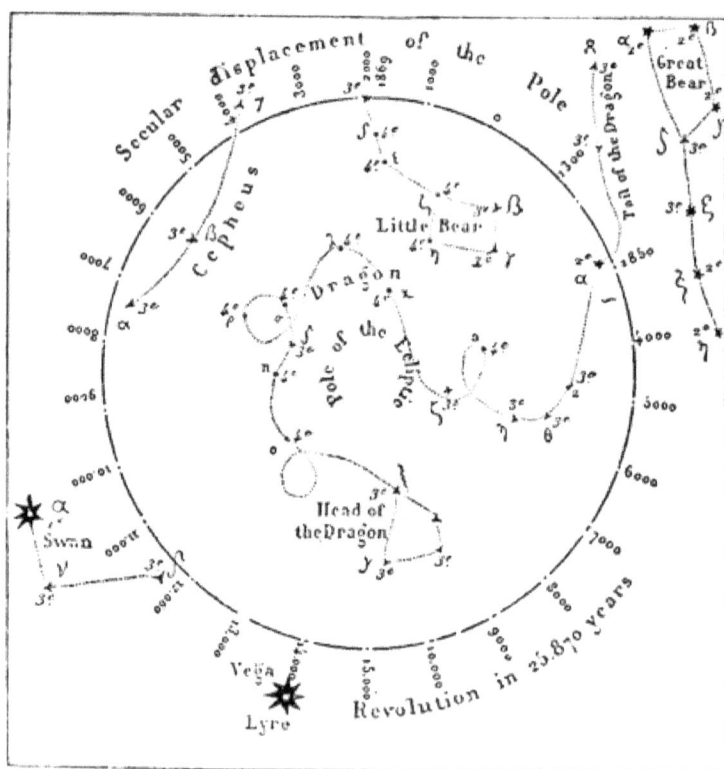

FIG. 12.

their poles, will change in the same way; the pole of the
equator, round which the heavens appear to move, describing
a curve about the pole of the ecliptic; and since the ecliptic

and equator are always *nearly* at the same angle, this curve
will be very nearly a circle, as represented on preceding page.

Now the pole of the equator is a very marked point in
the heavens, because the star nearest to it appears to have
no motion. If then we draw such a figure as above, so as
to see where this pole would be at any given date, and then
read in any old record that such and such a star had no
motion, we know at once at what date such a statement
must have been made. This means of estimating dates is
less certain than the other, because any star that is nearer to
the pole than any other will appear to have no motion
relatively to the rest, unless accurate measurements were
made. Nevertheless, when we have any reason to believe
that observations were carefully made, and there is any
evidence that some particular star was considered the Pole
Star, we have some confidence in concluding the date,
examples of which will appear in the sequel; and we may
give one illustration now, though not a very satisfactory one.
Hipparchus cites a passage from the sphere of Eudoxus, in
which he says, *Est vero stella quædam in eodem consistens
loco, quæ quidem polus est mundi.* (There is a certain immovable star, which is the pole of the world.)

Now referring to our figure, we find that about 1300 B.C.
the two stars, β Ursæ Minoris and κ Draconis were fairly
near the pole, and this fact leads us to date the invention of
this sphere at about this epoch, rather than a little before or

a little after, although, of course, there is nothing in *this*
argument (though there may be in others), to prevent us
dating it when *a* Draconis was near the pole, 2850 B.C.   This
star was indeed said by the Chinese astronomers in the reign
of Hoangti to mark the pole, which gives a date to their
observations.   The chief use of this latter method is to
confirm our conclusions from the former, rather than to
originate any.   Let us now apply our knowledge to the facts.

In the first place we may notice that in the time of
Hipparchus the vernal equinox was in the first degree of the
Ram, from which our own arrangement has originated.
Hipparchus lived 128 years B.C., or nearly 2,000 years ago, at
which time the equinox was exactly at β Aries.   Secondly, there
are many reasons for believing that at the time of the invention of the zodiac, indeed in the first dawning of
astronomy, the Bull was the first sign into which the sun
entered at the vernal equinox.   Now it takes 2,156 years to
retrograde through a sign, and therefore the Bull might occupy
this position any time between 2400 and 4456 B.C., and any
nearer approximation must depend on our ability to fix
on any particular *part* of the constellation as the original
equinoctial point.   We may say that whoever invented the
zodiac would no doubt make this point the *beginning* of a
sign, and therefore date its invention 2400 B.C. ; or on the
other hand, if it can be proved that the constellations were
known and observed before this, we may have to put back

the date to near the end of the sign, and make its last
remarkable stars the equinoctial ones, say those in the
horns of Taurus. Compare the line of Virgil,

> " Candidus auratis aperit cum cornibus annum
>   Taurus."

The date in this case would be about 4500 B.C.—or once
more some remarkable part of the constellation may give
proof that its appearance with the sun commenced the year
—and our date would be intermediate between these two.
In fact, the remarkable group of stars known as the *Pleiades*
actually does play this part. So much interest clusters, however, round this group, so much light is thrown by it on the
past history of astronomical ideas—and so much new information has recently been obtained about it—that it requires
a chapter to itself, and we shall therefore pass over its
discussion here. Let us now review some of the indications
that some part of the constellation of the Bull was originally
the first sign of the zodiac.

We need perhaps only mention the astrological books of
the Jews—the Cabal—in which the Bull is dealt with as the
first zodiacal sign. Among the Persians, who designate the
successive signs by the letters of the alphabet, *A* stands
for Taurus, *B* for the Twins, and so on. The Chinese attribute the commencement of the sun's apparent motion to the
stars of Taurus. In Thebes is a sepulchral chamber with

zodiacal signs, and Taurus at the head of them. The zodiac of the pagoda of Elephanta (Salsette) commences with the same constellation.

However, reasons have been given for assigning to the zodiac a still earlier date than this would involve. Thus Laplace writes :—" The names of the constellations of the zodiac have not been given to them by chance—they embody the results of a large number of researches and of astronomical systems. Some of the names appear to have reference to the motion of the sun. The Crab, for instance, and the He-Goat, indicate its retrogression at the solstices. The Balance marks the equality of the days and nights at the equinoxes, and the other names seem to refer to agriculture and to the climate of the country in which the zodiac was invented. The He-Goat appears better placed at the highest point of the sun's course than the lowest. In this position, which it occupied fifteen thousand years ago, the Balance was at the vernal equinox, and the zodiacal constellations match well with the climate and agriculture of Egypt." If we examine this, however, we see that all that is probable in it is satisfied by the Ram being at the vernal, and the Balance at the autumnal equinox, which corresponds much better with other evidence.

In the first instance, no doubt, the names of the zodiacal constellations would depend on the principal star or stars in each, and these stars and the portion of the ecliptic assigned

THE ZODIAC OF DENDERAH.

to each may have been noticed before the stars round them were grouped into constellations with different names. In any case, the introduction of the zodiac into Greece seems to have been subsequent to that of the celestial sphere, and not to have taken place more than five or six centuries before our era. Eudemus, of Rhodes, one of the most distinguished of the pupils of Aristotle, and author of a History of Astronomy, attributes the introduction of the zodiac to Œnopides of Chio, a contemporary of Anaxagoras. They did not receive it complete, as at first it had only eleven constellations, one of them, the Scorpion, being afterwards divided, to complete the necessary number. Their zodiacal divisions too would have been more regular had they derived them directly from the East, and would not have stretched in some instances over 36° to 48°, like the Lion, the Bull, the Fishes, or the Virgin—while the Crab, the Ram, and the He-Goat, have only 19° to 23°. Nor would their constellations be disposed so irregularly, some to the north and some to the south of the ecliptic, nor some spreading out widely and others crammed close together, so that we see that they only borrowed the idea from the Easterns, and filled it out with their ancient constellations. Such is the opinion of Humboldt.

With regard to the origin of the names of the signs of the zodiac, we must remember that a certain portion of the zodiacal circle, and not any definite group of stars, forms each sign, and that the constellations may have been formed

separately, and have received independent names, though
afterwards receiving those of the sign in which they were.
The only rational suggestion for the origin of the names is
that they were connected with some events which took place,
or some character of the sun's motion observed, when it was
in each sign. Thus we have seen that the Balance may refer
to equal nights and days (though only introduced among
the *Greeks* in the time of Hipparchus), and the Crab to the
retrogression or stopping of the sun at the solstice.

The various pursuits of husbandry, having all their
necessary times, which in the primeval days were determined
by the positions of the stars, would give rise to more
important names. Thus the Ethiopian, at Thebes, would call
the stars that by their rising at a particular time indicated
the inundation, Aquarius, or the Waterer; those beneath
which it was necessary to put the plough to the earth, the
Bull stars. The Lion stars would be those at whose appearance this formidable animal, driven from the deserts by
thirst, showed himself on the borders of the river. Those
of the Ear of Corn, or the Virgin of harvest, those beneath
which the harvest was to be gathered in; and the sign of
the Goat, that in which the sun was when these animals
were born.

There can be but little doubt but that such was the origin
of the names imposed, and for a time they would be understood in that sense. But afterwards, when time was more

accurately kept, and calendars regulated, without each man
studying the stars for himself, when the precession of the
equinoxes made the periods not exactly coincide, the
original meaning would be lost, the stars would be associated
with the animals, as though there was a real bull, a real lion,
&c., in the heavens; and then the step would be easy to
represent these by living animals, whom they would endow
with the heavenly attributes of what they represented; and
so the people came at last to pray to and worship the several
creatures for the sake of their supposed influence. They
asked of the Ram from their flocks the influences they
thought depended on the constellation. They prayed the
Scorpion not to spread his evil venom on the world; they
revered the Crab, the Scarabæus, and the Fish, without
perceiving the absurdity of it.

It is certain at least that the gods of many nations are
connected or are identical with the signs of the zodiac, and
it seems at least more reasonable to suppose the former
derived from the latter than *vice versá.*

Among the Greeks indeed, who had, so to speak, their
gods ready made before they borrowed the idea of the zodiac,
the process appears to have been the reverse, they made the
signs to represent as far as they could their gods. In the
more pastoral peoples, however, of the East, and in Egypt, this
process can be very clearly traced. Among the Jews there
seems to be some remarkable connection between their

patriarchs and these signs, though the history of that connection may not well be made out. The twelve signs are
mentioned as being worshipped, along with the sun and
moon, in the Book of Kings. But what is more remarkable
is the dream of Joseph, in which the sun and moon and the
other eleven stars worshipped him, coupled with the various
designations or descriptions given to each son in the blessing
of Jacob. In Reuben we have the man who is said to be
"unstable as water," in which we may recognise Aquarius.
In Simeon and Levi "the brethren," we trace the Twins.
Judah is the "Lion." Zebulun, "that dwells at the haven
of the sea," represents Fishes. Issachar is the Bull, or
"strong ass couching down between two burdens." Dan,
"the serpent by the way, the adder in the path," represents
the Scorpion. Gad is the Ram, the leader to a flock or
troop of sheep. Asher the Balance, as the weigher of
bread. Naphtali, "the hind let loose," is the Capricorn,
Joseph the Archer, whose bow abode in strength. Bru-
jamin the Crab, changing from morning to evening, and
Dinah, the only daughter, represents the Virgin.

There is doubtless something far-fetched in some of these
comparisons, but when we consider the care with which the
number twelve was retained, and that the four chief tribes
carried on their sacred standards these very signs—namely,
Judah a lion, Reuben a man, Ephraim a bull, and Dan a scorpion—and notice the numerous traces of astronomical culture

in the Jewish ceremonies, the seven lights of the candlestick, the twelve stones of the High Priest, the feasts at the two equinoxes, the ceremonies connected with a ram and a bull, we cannot doubt that there is something more than chance in the matter, but rather conclude that we have an example of the process by which, in the hands of the Egyptians themselves, astronomical representations became at last actually deified.

It has been thought possible indeed to assign definitely each god of the Egyptians to one of the twelve zodiacal signs. The Ram was consecrated to Jupiter Ammon, who was represented with a ram's head and horns. The Bull became the god Apis, who was worshipped under that similitude. The Twins correspond to Horus and Harpocrates, two sons of Osiris. The Crab was consecrated to Anubis or Mercury. The Lion belonged to the summer sun, Osiris; the Virgin to Isis. The Balance and the Scorpion were included together under the name of Scorpion, which animal belongs to Typhon, as did all dangerous animals. The Archer was the image of Hercules, for whom the Egyptians had great veneration. The Capricorn was consecrated to Pan or Mendes. The Waterer—or man carrying a water-pot —is found on many Egyptian monuments.

This process of deification was rendered easier by the custom they had of celebrating a festival each month, under the name *neomenia*. They characterised the neomenias of

the various months by making the animal whose sign the
sun was entering accompany the Isis which announced
the *fête*. They were not content with a representation only,
but had the animal itself. The dog, being the symbol of
Canuulus, with which the year commenced, a living dog
was made to head the ceremonial of the first neomenia.
Diodorus testifies to this as an eye-witness.

These neomenias thus came to be called the festival of
the Bull, of the Ram, the Dog, or the Lion. That of the
Ram would be the most solemn and important in places
where they dealt much in sheep. That of the Bull in the
fat pasture-lands of Memphis and Lower Egypt. That of
Capella would be brilliant at Mendes, where they bred goats
more than elsewhere.

We may fortify these opinions by a quotation from Lucian,
who gives expression to them very clearly. " It is from the
divisions of the zodiac," he says, " that the crowd of animals
worshipped in Egypt have had their origin. Some employed
one constellation, and some another. Those who used to
consult that of the Ram came to adore a ram. Those who
took their presages from the Fishes would not eat fish. The
goat was not killed in places were they observed Capricornus,
and so on, according to the stars whose influence they cared
most for. If they adored a bull it was certainly to do
honour to the celestial Bull. The Apis, which was a sacred
object with them, and wandered at liberty through the

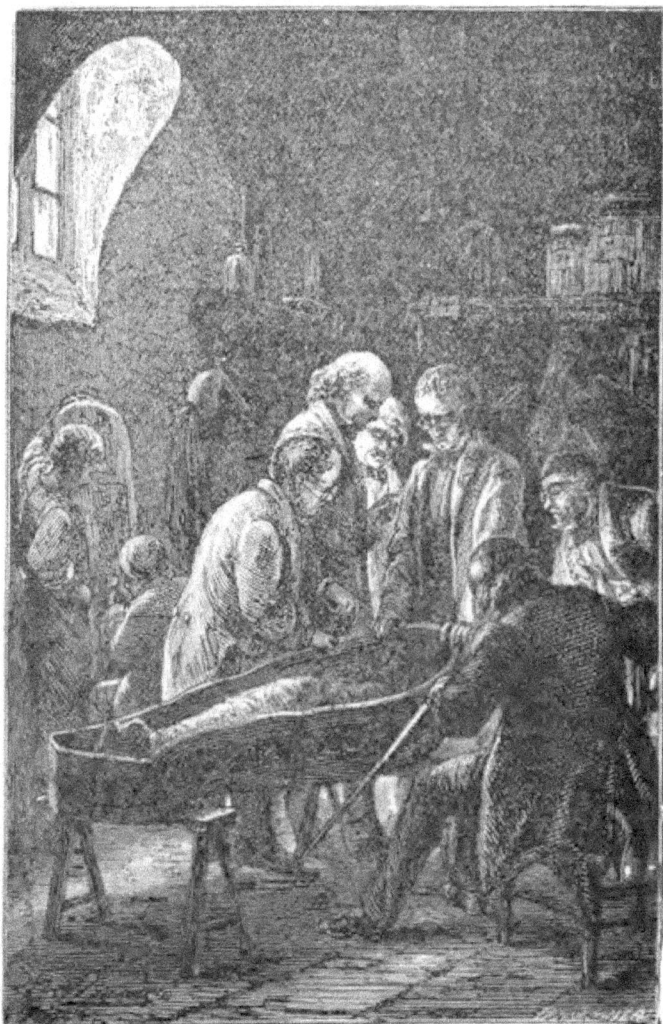

PLATE IV. —THE ZODIAC AND THE DEAD IN EGYPT.

country, and for which they founded an oracle, was the astrological symbol of the Bull that shone in the heavens."

Their use of the zodiac is illustrated in an interesting manner by a mummy found some years ago in Egypt. At the bottom of the coffin was found painted a zodiac, something like that of Denderah; underneath the lid, along the body of a great goddess, were drawn eleven signs, but with that of *Capricornus* left out. The inscription showed that the mummy was that of a young man, aged 21 years, 4 months, and 22 days, who died the 19th year of Trajan, on the 8th of the month Pazni, which corresponds to the 2nd of June, A.D. 116. The embalmed was therefore born on the 12th of January, A.D. 95, at which time the sun was in the constellation of Capricornus. This shows that the zodiac was the representation of the astrological theories about the person embalmed, who was doubtless a person of some importance. (See Plate IV.) Any such use as this, however, must have been long subsequent to the invention of the signs themselves, as it involves a much more complicated idea.

CHAPTER V.

AMONG the most remarkable of the constellations is a group of seven stars arranged in a kind of triangular cluster, and known as the Pleiades. It is not, strictly speaking, one of the constellations, as it forms only part of one. We have seen that one of the ancient signs of the zodiac is the Bull, or Taurus; the group of stars we are now speaking of forms part of this, lying towards the eastern part in the shoulders of the Bull. The Pleiades scarcely escape anybody's observation now, and we shall not be, therefore, surprised that they have always attracted great attention. So great indeed has been the attention paid to them that festivals and seasons, calendars and years, have by many nations been regulated by their rising or culmination, and they have been thus more mixed up with the early history of astronomy, and have left more marks on the records of past nations, than any other celestial object, except the sun and moon.

The interesting details of the history of the Pleiades have been very carefully worked out by R. G. Haliburton, F.S.A., to whom we owe the greater part of the information we possess on the subject.[1]

Let us first explain what may be observed with respect to the Pleiades. It is a group possessing peculiar advantages for observation; it is a compact group, the whole will appear at once; and it is an unmistakable group and it is near the equator, and is therefore visible to observers in either hemisphere.

Now suppose the sun to be in the same latitude as the Pleiades on some particular day; owing to the proximity of the group to the ecliptic, it will be then very near the sun, and it will set with it and be invisible during the night. If the sun were to the east of the Pleiades they would have already set, and the first view of the heavens at sunset would not contain this constellation; and so it would be so long as the sun was to the east, or for nearly half a year; though during some portion of this time it would rise later on in the night. During the other half year, while the sun was to the west, the Pleiades would be visible at sunset, and we immediately see how they are thus led to divide the whole year into two

---

[1] Mr. Haliburton's observations are contained in an interesting pamphlet, entitled *New Materials for the History of Man*, which is quoted by Prof. Piazzi Smyth, but which is not easy to obtain. It may be seen, however, in the British Museum.

portions, one of which might be called *the Pleiades below*, and the other *the Pleiades above*. It is plain that the Pleiades first become visible at sunset, when they are then just rising, in which case they will culminate a little after midnight (not at midnight, on account of the twilight) and be visible all night. This will occur when the sun is about half a circle removed from them—that is, at this time, about the beginning of November; which would thus be the commencement of one half of the year, the other half commencing in May. The culmination of the Pleiades at midnight takes place a few days later, when they rise at the time that the sun is really on the horizon, in which case they are exactly opposite to it; and this will happen on the same day all over the earth. The opposite effect to this would be when the sun was close to the Pleiades—a few days before which the latter would be just setting after sunset, and a few days after would be just rising before sunrise.

We have thus the following observations, that might be made with respect to this, or any other well-marked constellation. First, the period during which it was visible at sunset; secondly, the date of its culmination at midnight; thirdly, its setting in the evening ; and fourthly, its rising in the morning : the last two dates being nearly six months removed from the second. There are also the dates of its culmination at sunrise and sunset, which would divide these intervals into two equal halves. On account of the preces-

I

sion of the equinoxes, as explained in the last chapter, the time at which the sun has any particular position with respect to the stars, grows later year by year in relation to the equinoctial points. And as we regulate our year by the date of the sun's entrance on the northern hemisphere, the sidereal dates, as we may call them, keep advancing on the months. As, however, the change is slow, it has not prevented years being commenced and husbandry being regulated by the dates above mentioned. Any date that is regulated by the stars we might expect to be nearly the same all over the world, and the customs observed to be universal, though the date itself might alter, and in this way. So long as the date was directly obtained from the position of the star, all would agree ; but as soon as a solar calendar was arranged, and it was found that at that time this position coincided with a certain day, say the Pleiades culminating at midnight on November 17, then some would keep on the date November 17 as the important day, even when the Pleiades no longer culminated at midnight then, and others would keep reckoning by the stars, and so have a different date.

With these explanations we shall be able to recognise how much the configurations of the Pleiades have had to do with the festivals and calendars of nations, and have even left their traces on customs and names in use among ourselves to the present day.

We have evidences from two very different quarters of the universality of the division of the year into two parts by means of the Pleiades. On the one hand we learn from Hesiod that the Greeks commenced their winter seasons in his days by the setting of the Pleiades in the morning, and the summer season by their rising at that time. And Mr. Ellis, in his *Polynesian Researches*, tells us that "the Society Islanders divided the year into two seasons of the Pleiades, or *Matarii*. The first they called *Matarii i nia*, or the *Pleiades above*. It commenced when, in the evening, these stars appeared at or near the horizon, and the half year during which, immediately after sunset, they were seen above the horizon was called *Matarii i nia*. The other season commenced when at sunset these stars are invisible, and continued until at that time they appeared again above the horizon. This season was called *Matarii i raro*, i.e. *the Pleiades below*." Besides these direct evidences we shall find that many semi-annual festivals connected with these stars indicate the commencement of the two seasons among other nations.

One of these festivals was of course always taken for the commencement of the year, and much was made of it as new-year's day. A new-year's festival connected with and determined by the Pleiades appears to be one of the most universal of all customs ; and though some little difficulty arises, as we have already pointed out, in fixing the date with

I 2

reference to solar calendars, and differences and coincidences
in this respect among different nations may be to a certain
extent accidental, yet the fact of the wide-spread observance
of such a festival is certain and most interesting.

The actual observance at the present day of this festival
is to be found among the Australian savages. At their
midnight culmination in November, they still hold a new-
year's *corroboree*, in honour of the *Mormodellick*, as they
call the Pleiades, which they say are "very good to the
black fellows." With them November is somewhat after the
beginning of spring, but in former days it would mark the
actual commencement, and the new year would be regulated
by the seasons.

In the northern hemisphere this culmination of the Pleiades
has the same relation to the autumnal equinox, which would
never be taken as the commencement of the year; and we
must therefore look to the southern hemisphere for the origin
of the custom ; especially as we find the very Pleiades themselves called *Vergiliæ*, or stars of spring. Of course we might
suppose that the rising of the constellation in the *morning*
had been observed in the northern hemisphere, which would
certainly have taken place in the beginning of spring some
5,000 years ago ; but this seems improbable, first, because it
is unlikely that different phenomena of the Pleiades should
have been most noticed, and secondly, because neither April
nor May are among any nations connected with this constel-

lation by name. Whereas in India the year commenced in
the month they called *Cartiguey,* which means the Pleiades.
Among the ancient Egyptians we find the same connection
between *Athar-aye,* the name of the Pleiades, with the
Chaldeans and Hebrews, and *Athor* in the Egyptian
name of November. The Arabs also call the constellation
Atauria. We shall have more to say on this etymology
presently, but in the meantime we learn that it was the
phenomenon connected with the Pleiades at or about
November that was noticed by all ancient nations, from
which we must conclude that the origin of the new-year's
spring festival came from the southern hemisphere.

There is some corroboration of this in the ancient traditions
as to the stars having changed their courses. In the
southern hemisphere a man standing facing the position of
the sun at noon would see the stars rise on his right hand
and move towards his left. In the northern hemisphere, if
he also looked in the direction of the sun at noon, he would
see them rise on his left hand. Now one of a race migrating
from one side to the other of the equator would take his
position from the sun, and fancy he was facing the same way
when he looked at it at noon, and so would think the
motion of the stars to have altered, instead of his having
turned round. Such a tradition, then, seems to have arisen
from such a migration, the fact of which seems to be confirmed
by the calling the Pleiades stars of spring, and commencing

the year with their culmination at midnight. In order to trace this new-year's festival into other countries, and by this means to show its connection with the Pleiades, we must remark that every festival has its peculiar features and rites, and it is by these that we must recognise it, where the actual date of its occurrence has slightly changed ; bearing, of course, in mind that the actual change of date must not be too great to be accounted for by the precession of the equinoxes, or about seventy-one years for each day of change, since the institution of the festival, and that the change is in the right direction.

Now we find that everywhere this festival of the Pleiades' culmination at midnight (or it may be of the slightly earlier one of their first appearance at the horizon at apparent sunset) was always connected with the memory of the dead. It was a " feast of ancestors."

Among the Australians themselves, the *corroborees* of the natives are connected with a worship of the dead. They paint a white stripe over their arms, legs, and ribs, and, dancing by the light of their fires by night, appear like so many skeletons rejoicing. What is also to be remarked, the festival lasts three days, and commences in the evening; the latter a natural result of the date depending on the appearance of the Pleiades on the horizon at that time.

The Society Islanders, who, as we have seen, divided their year by the appearance of the Pleiades at sunset, commenced their year on the first day of the appearance, about

November, and also celebrated the closing of one and the opening of a new year by a " usage resembling much the popish custom of mass for souls in purgatory," each man returning to his home to offer special prayers for the spirits of departed relatives.

In the Tonga Islands, which belong to the Fiji group, the festival of *Inachi*, a vernal first-fruits' celebration, and also a commemoration of the dead takes place towards the end of October, and commences at sunset.

In Peru the new-year's festival occurs in the beginning of November, and is " called *Ayamarca* from *aya*, a corpse, and *marca*, carrying in arms, because they celebrated the solemn festival of the dead, with tears, lugubrious songs, and plaintive music; and it was customary to visit the tombs of relations, and to leave in them food and drink." The fact that this took place at the time of the discovery of Peru on the very same day as a similar ceremony takes place in Europe, was only an accidental coincidence, which is all the more remarkable because the two appear, as will be seen in the sequel, to have had the same origin, and therefore at first the same date, and to have altered from it by exactly the same amount. These instances from races south of the equator prove clearly that there exists a very general connection with new-year's day, as determined by the rising of the Pleiades at sunset, and a festival of the dead; and in some instances with an offering of first-fruits. What the origin of this

connection may be is a more difficult matter. At first sight
one might conjecture that with the year that was passed it
was natural to connect the men that had passed away ; and
this may indeed be the true interpretation : but there are
traditions and observances which may be thought by some to
point to some ancient wide-spread catastrophe which happened
at this particular season, which they yearly commemorated,
and reckoned a new year from each commemoration. Such
traditions and observances we shall notice as we trace the
spread of this new-year's festival of the dead among various
nations, and its connection with the Pleiades.

We have seen that in India November is called the month
of the Pleiades. Now on the 17th day of that month is
celebrated the Hindoo Durga, a festival of the dead, and said
by Greswell to have been a new-year's commemoration at the
earliest time to which Indian calendars can be carried back.

Among the ancient Egyptians the same day was very
noticeable, and they took care to regulate their solar calendars
that it might remain unchanged. Numerous altered calendars
have been discovered, but they are all regulated by this one
day. This was determined by the culmination of the Pleiades
at midnight. On this day commenced the solemn festival of
the Isia, which, like the *corroborees* of the Australians, lasted
three days, and was celebrated in honour of the dead, and
of Osiris, the lord of tombs. Now the month Athyr was un-
doubtedly connected with the Pleiades, being that "in which

the Pleiades are most distinct"—that is, in which they rise near and before sunset. Among the Egyptians, however, more attention was paid to astronomy than amongst the savage races with which the year of the Pleiades would appear to have originated, and they studied very carefully the connection between the positions of the stars and the entrance of the sun into the northern hemisphere, and regulated their calendar accordingly; as we shall see shortly in speaking of the pyramid builders.

The Persians formerly called the month of November *Mordád*, the angel of death, and the feast of the dead took place at the same time as in Peru, and was considered a new-year's festival. It commenced also in the evening.

In Ceylon a combined festival of agriculture and of the dead takes place at the beginning of November.

Among the better known of the ancient nations of the northern hemispheres, such as the Greeks and Romans, the anomaly of having the beginning of the year at the autumnal equinox seems to have induced them to make a change to that of spring, and with this change has followed the festival of the dead, although some traces of it were left in November.

The commemoration of the dead was connected among the Egyptians with a deluge, which was typified by the priest placing the image of Osiris in a sacred coffer or ark, and launching it out into the sea till it was borne out of sight.

Now when we connect this fact, and the celebration taking place on the 17th day of Athyr, with the date on which the Mosaic account of the deluge of Noah states it to have commenced, " in the second month (of the Jewish year, which corresponds to November), the 17th day of the month," it must be acknowledged that this is no chance coincidence, and that the precise date here stated must have been regulated by the Pleiades, as was the Egyptian date. This coincidence is rendered even stronger by the similiarity of traditions among the two nations concerning the dove and the tree as connected with the deluge. We find, however, no festival of the dead among the Hebrews; their better form of faith having prevented it.

We have not as yet learnt anything of the importance of the Pleiades among the ancient Babylonian astronomers, but as through their tablets we have lately become acquainted with their version of the story of the deluge, we may be led in this way to further information about their astronomical appreciation of this constellation.

From whatever source derived, it is certain that the Celtic races were partakers in this general culture, we might almost call it, of the Pleiades, as shown by the time and character of their festival of the dead. This is especially interesting to ourselves, as it points to the origin of the superstitions of the Druids, and accounts for customs remaining even to this day amongst us.

PLATE V.—THE LEGENDS OF THE DRUIDS.

The first of November was with the Druids a night full of
mystery, in which they annually celebrated the reconstruction of the world.   A terrible rite was connected with this;
for the Druidess nuns were obliged at this time to pull down
and rebuild each year the roof of their temple, as a symbol
of the destruction and renovation of the world.   If one of
them, in bringing the materials for the new roof, let fall her
sacred burden, she was lost.   Her companions, seized with a
fanatic transport, rushed upon her and tore her to pieces, and
scarcely a year is said to have passed without there being one
or more victims.   On this same night the Druids extinguished
the sacred fire, which was kept continually burning in the
sacred precincts, and at that signal all the fires in the island
were one by one put out, and a primitive night reigned
throughout the land.   Then passed along to the west the
phantoms of those who had died during the preceding year,
and were carried away by boats to the judgment-seat of the god
of the dead.   (Plate V.)   Although Druidism is now extinct,
the relics of it remain to this day, for in our calendar we still
find November 1 marked as All Saints' Day, and in the pre-
Reformation calendars the last day of October was marked
All Hallow Eve, and the 2nd of November as All Souls';
indicating clearly a three days' festival of the dead, commencing in the evening, and originally regulated by the Pleiades—
an emphatic testimony how much astronomy has been
mixed up with the rites and customs even of the English of

to-day. In former days the relics were more numerous, in the Hallowe'en torches of the Irish, the bonfires of the Scotch, the *coel-coeth* fires of the Welsh, and the *tindle* fires of Cornwall, all lighted on Hallowe'en. In France it still lingers more than here, for to this very day the Parisians at this festival repair to the cemeteries, and lunch at the graves of their ancestors.

If the extreme antiquity of a rite can be gathered from the remoteness of the races that still perform it, the fact related to us by Prescott in his *History of the Conquest of Mexico* cannot fail to have great interest. There we find that the great festival of the Mexican cycle was held in November, at the time of the midnight culmination of the Pleiades. It began at sunset, and at midnight as that constellation approached the zenith, a human victim, was offered up, to avert the dread calamity which they believed impended over the human race. They had a tradition that the world had been previously destroyed at this time, and they were filled with gloom and dismay, and were not at rest until the Pleiades were seen to culminate, and a new cycle had begun; this great cycle, however, was only accomplished in fifty-two years.

It is possible that the festival of lanthorns among the Japanese, which is celebrated about November, may be also connected with this same day, as it is certain that that nation does reckon days by the Pleiades.

These instances of a similar festival at approximately the

same period of the year, and regulated (until fixed to a
particular day in a solar calendar) by the midnight culmination of the Pleiades, show conclusively how great an influence
that constellation has had on the manners and customs of the
world, and throw some light on the history of man.

Even where we find no festival connected with the particular
position of the Pleiades which is the basis of the above, they
still are used for the regulation of the seasons—as amongst
the Dyaks of Borneo.   This race of men are guided in their
farming operations by this constellation.   "When it is low in
the east at early morning, before sunrise, the elders know it
is time to cut down the jungle; when it approaches mid-
heaven, then it is time to burn what they have cut down;
when it is declining towards the west, then they plant; and
when in the early evening it is seen thus declining, then they
may reap in safety and in peace;" the latter period is also
that of their feast of *Nycapian*, or first-fruits.

We find the same regulations amongst the ancient Greeks
in the days of Hesiod, who tells us that the corn is to be cut
when the Pleiades rise, and ploughing is to be done when they
set.   Also that they are invisible for forty days, and reappear
again at harvest.   When the Pleiades rise, the care of the vine
must cease; and when, fleeing from Orion, they are lost in the
waves, sailing commences to be dangerous.   The name, indeed,
by which we now know these stars is supposed to be derived
from the word πλεῖν, to sail—because sailing was safe after

they had risen; though others derive it from πέλειαι, a flight of doves.

Any year that is regulated by the Pleiades, or by any other group of stars, must, as we have seen before, be what is called a sidereal, and not a solar year. Now a year in un-civilised countries can only mean a succession of seasons, as is illustrated by the use of the expression "a person of so many summers." It is difficult of course to say when any particular season begins by noticing its characteristics as to weather; even the most regular phenomena are not certain enough for that; we cannot say that when the days and nights become exactly equal any marked change·takes place in the temperature or humidity of the atmosphere, or in any other easily-noticed phenomena. The day therefore on which spring commences is arbitrary, except that, inasmuch as spring depends on the position of the sun, its commencement, ought to be regulated by that luminary, and not by some star-group which has no influence in the matter. Nevertheless the position of such a group is much more easily observed, and in early ages could almost alone be observed ; and so long as the midnight culmination of the Pleiades—judged of, it must be noticed, by their appearance *on the horizon* at sun-set—fairly coincided with that state of weather which might be reckoned the commencement of spring conditions, no error would be detected, because the change in their position is so slow. The solar spring is probably a later discovery, which now, from its greater reasonableness and constancy, has superseded the old one. But since the time of the sun's crossing the equator is the natural commencement of spring, whether discovered or not, it is plain that no group of stars could be taken as a guide instead, if their indication did not approximately coincide with this.

If then we can determine the exact date at which the Pleiades indicated by their midnight culmination the sun's passage across the equator, we can be sure that the spring could only have been regulated by this during, say, a thousand years at most, on either side of this date. It is very certain that if the method of reckoning spring by the stars had been invented at a more remote date, some other set of stars would have been chosen instead.

Now when was this date? It is a matter admitting of certain calculation, depending only on numbers derived from observation in our own days and records of the past few centuries, and the answer is that this date is about 2170 B.C.

We have seen that, though it was probably brought from the southern hemisphere, the Egyptians adopted the year of the Pleiades, and celebrated the new-year's festival of the dead ; but they were also advanced astronomers, and would soon find out the change that took place in the seasons when regulated by the stars. And to such persons the date at which the two periods coincided, or at least were exactly half a year apart, would be one of great importance and interest, and

there seems to be evidence that they did commemorate it in
a very remarkable manner. The evidence, however, is all
circumstantial, and the conclusion therefore can only claim
probability. The evidence is as follows:—The most remarkable buildings of Egypt are the pyramids. These are of
various sizes and importance, but are built very much after
the same plan. They seem, however, to be all copies from
one, the largest, namely, the Pyramid of Gizeh, and to be of
subsequent date to this. Their object has long been a puzzle,
and the best conclusion has been supposed to be that they
were for sepulchral purposes, as in some of them coffins
have been found. The large one, however, shows far more
than the rest of the structure, and cannot have been meant
for a funeral pile alone.

Its peculiarities come out on a careful examination and
measurement such as it has been subjected to at the devoted
hands of Piazzi Smyth, the Astronomer Royal for Scotland.
He has shown that it is not built at random, as a tomb might
be, but it is adjusted with exquisite design, and with surprising accuracy. In the first place it lies due north, south, east,
and west, and the careful ascertainment of the meridian of the
place, by modern astronomical instruments, could not suggest
any improvement in its position in this respect. The outside
of it is now, so to speak, pealed, that is to say, there was
originally, covering the whole, another layer of stones which
have been taken away. These stones, which were of a dif-

K

ferent material, were beautifully polished, as some of the
remaining ones, now covered and concealed, can testify. The
angle at which they are cut, and which of course gives the
angle and elevation of the whole pyramid, is such that the
height of it is in the same proportion to its circumference or
perimeter, as the radius of a circle is to its circumference
approximately. The height, in fact, is proved by measurement
and observation to be 486 ft., and the four sides together to be
3,056 ft., or about 6¼ times the height. It does not seem improbable that, considering their advancement, the Egyptians
might have calculated approximately how much larger the
circumference of the circle is than its diameter, and it is a
curious coincidence that the pyramid expresses it. Professor
Piazzi Smyth goes much further and believes that they knew,
or were divinely taught, the shape and size of the earth, and
by a little manipulation of the length of their unit, or as he
expresses it the " pyramid inch," he makes the base of the
pyramid express the number of miles in the diameter of the
earth.

Now in the interior of the apparently solid structure,
besides the usual slanting passage down to a kind of cellar or
vault beneath the middle of the base, which may have been
used for a sepulchral resting-place, there are two slanting
passages, one running north and the other running south, and
slanting up at different angles. Part of that which leads
south is much enlarged, and is known as the grand gallery.

It is of a very remarkable shape, being perfectly smooth and
polished along its ascending base, as indeed it is in every
part, and having a number of steps or projections, pointing
also upwards at certain angles, very carefully maintained.
Whether we understand its use or not, it is very plain that
it has been made with a very particular design, and one
not easily comprehended. This leads into a chamber
known as the king's chamber, whose walls are exquisitely
polished and which contains a coffer known as *Cheops' Coffin.*
This coffer has been villainously treated by travellers, who
have chipped and damaged it, but originally it was very
carefully made and polished. It is too large to have been
brought in by the only entrance into the chamber after it was
finished, and therefore is obviously no coffin at all, as is proved
also by the elaborateness of the means of approach. Professor
Piazzi Smyth has made the happy suggestion that it represents
their standard of length and capacity, and points out the
remarkable fact that it contains exactly as much as four
quarters of our dry measure. As no one has ever suggested
what our "quarters" are quarters of, Professor Smyth very
naturally supplies the answer—"of the contents of the
pyramid coffer." There are various other measurements that
have been made by the same worker, and their meaning suggested in his interesting book, *Our Inheritance in the Great
Pyramid,* which we may follow or agree to as we can; but
from all that has been said above, it will appear probable that

this pyramid was built with a definite design to mark various natural phenomena or artificial measures, which is all we require for our present purpose. Now we come to the question, what is the meaning of the particular angles at which the north-looking and south-looking passages rise, if, as we now believe, they must have *some* meaning.

The exits of these passages were closed, and they could not therefore have been for observation, but they may have been so arranged as to be a memorial of any remarkable phenomena to be seen in those directions. To ascertain if there be any such to which they point, we must throw back the heavens to their position in the days of the Egyptians, because, as we have seen, the precession of the equinoxes alters the meridian altitude of every star. As the passages point north and south, if they refer to any star at all, it must be to their passing the meridian.

Now let us take the heavens as they were 2170 B.C., the date at which the Pleiades *really* commenced the spring, by their midnight culmination, and ask how high they would be then. The answer of astronomy is remarkable—*Exactly at that height that they could be seen in the direction of the southward-pointing passage of the pyramid.*" And would any star then be in a position to be seen in the direction of the other or northward-looking passage? Yes, the largest star in the constellation of the Dragon, which would be so near the pole (3° 52') as to be taken as the Pole Star in those days.

These are such remarkable coincidences in a structure admittedly made with mathematical accuracy and design, and truly executed, that we cannot take them to be accidental, but must endeavour to account for them.

The simplest explanation seems to be, that everything in the pyramid is intended to represent some standard or measure, and that these passages have to do with their year. They had received the year of the Pleiades from a remoter antiquity than their own, they had discovered the true commencement of solar spring, as determined from the solar autumnal equinox, and they commemorated by the building of the pyramid the coincidence of the two dates, making passages in it which would have no meaning except at that particular time.

Whether the pyramid was built *at that time*, or whether their astronomical knowledge was sufficient to enable them to predict it and build accordingly, just as we calculate back to it, we have no means of knowing. It is very possible that the pyramid may have been built by some immigrating race more learned in astronomy, like the Accadians among the Babylonians.

Either the whole of the conclusions respecting the pyramid is founded on pure imagination and the whole work upon it thrown away, or we have here another very remarkable proof of the influence of the Pleiades on the reckoning of the year, and a very interesting chapter in the history of the heavens.

Following the guidance of Mr. Haliburton, we shall find still more customs, and names depending in all probability on the influence the Pleiades once exerted, and the observances connected with the feasts in their honour.

The name by which the Pleiades are known among the Polynesians is the " Tau," which means a season, and they speak of the years of the Tau, that is of the Pleiades. Now we have seen that the Egyptians had similar feasts at similar times, in relation to this constellation, and argued that they did not arise independently. This seems still further proved by their name for these stars—the Atauria.

Now the Egyptians do not appear to have derived their signs of the Zodiac from the same source ; these had a Babylonian origin, and the constellation in which the Pleiades were placed by the latter people was the Bull, by whatever name he went. The Egyptians, we may make the fair surmise, adopted from both sources ; they took the Pleiades to indicate the Bull, and they called this animal after the Atauria. From thence we got the Latin Taurus, and the German Thier.

It is possible that this somehow got connected with the letter " tau " in Greek, which seems itself connected with the sacred scarabæus or Tau-beetle of Egypt ; but the nature of the connection is by no means obvious. Mr. Haliburton even suggests that the " tors " and " Arthur's seat," which are names given to British hill-tops, may be connected with the

"high places," of the worship of the Pleiades, but of this we
have no proof.

Among the customs possibly derived from the ancients,
through the Phœnicians, though now adopted as conveying a
different meaning in a Christian sense, is that of the "hot cross
bun," or "bull cake." It is found on Egyptian monuments,
signifying the four quarters of the year, and sometimes
stamped with the head and horns of the bull. It is found
among ourselves too, essentially connected with the dead, and
something similar to it appears in the "soul cake" connected
originally with All Souls' Day.

Among the Scotch it was traditionally thought that on New
Year's Eve the Candlemas Bull can be seen, rising at twilight
and sailing over the heavens—a very near approach to a matter-
of-fact statement.

We have seen that among the ancient Indians there was
some notice taken of the Pleiades, and that they in all probability guided their year by them or by some other stars: it
would therefore behove them to know something of the precession of the equinoxes. It seems very well proved that their
days of Brahma and other periods were meant to represent
some astronomical cycles, and among these we find one that is
applicable to the above. They said that in every thousand
divine ages, or in every day of Brahma, fourteen Menus are
successively invested with the sovereignty of the earth.
Each Menu transmits his empire to his sons during

seventy-one divine ages. We may find a meaning for this
by putting it that the equinox goes forward fourteen days
in each thousand years, and each day takes up seventy-one
years.

These may not be the only ones among the various customs,
sayings, and names that are due in one way or other to this
primitive method of arranging the seasons by the positions
of the stars, especially of those most remarkable and conspicuous ones the Pleiades, but they are those that are best
authenticated. If the connection between the Pleiades and
the festival of the dead, the new year and a deluge, can be
clearly made out; if the tradition of the latter be found
as universal as that of the former, and be connected with it
in the Mosaic narrative ; if we can trace all these traditions
to the south of the equator, and find numerous further traditions connected with islands, we may find some reason for
believing in their theory who suggest that the early progeni-
tors of the human race (? all of them) were inhabitants of
some fortunate islands of even temperature in the southern
hemisphere, where they made some progress in civilisation,
but that their island was swallowed up by the sea, and that
they only escaped by making huge vessels, and, being carried
by the waves, they landed on continental shores, where they
commemorated yearly the great catastrophe that had happened to them, notifying its time by the position of the
Pleiades, making it a feast of the dead whom they had left

behind, and opening the year with the day, whether it were
spring or not, and handing down to their descendants and to
those among whom they came, the traditions and customs
which such events had impressed upon them.

Whether such an account be probable, mythical, or unnatural, there are certainly some strange things to account for
in connection with the Pleiades.

CHAPTER VI.

MANY and various have been the ideas entertained by reflecting men in former times on the nature and construction of the heavenly vault, wherein appeared those stars and constellations whose history we have already traced. Is it solid? or liquid? or gaseous? Each of these and many other suppositions have been duly formulated by the ancient philosophers and sages, although, as we are told by modern astronomy, it does not exist at all.

In our study of the ancient ideas about the structure of the universe, we will commence with that early and curious system which considered the heavenly vault to be material and solid.

The theory of a solid sky received the assent of all the most ancient philosophers. In his commentary on Aristotle's work on the heavens, Simplicius reveals the repugnance the ancient philosophers felt in admitting that a star could stand alone in space, or have a free motion of its own. It must have a support, and they therefore conceived that the sky must be solid. However strange this idea may now appear, it formed for many centuries the basis of all astronomical theories. Thus Anaximenas (in the sixth century B.C.) is related by Plutarch to have said that "the outer sky is solid and crystalline," and that the stars are "fixed to its surface like studs," but he does not say on what this opinion was founded, though it is probable that, like his master Anaximander, he could not understand how the stars could move without being supported.

Pythagoras, who lived about the same epoch, is also supposed by some to have held the same views, and it is possible that they all borrowed these ideas from the Persians, whose earliest astronomers are said in the *Zend avesta* to have believed in concentric solid skies.

Eudoxus of Cnidus, in the fifth century B.C., is said by his commentator Aratus to have also believed in the solidity of the heavens, but his reasons are not assigned.

Notwithstanding these previously expressed opinions, Aristotle (fourth century, B.C.) has for a long time been generally supposed to be the inventor of solid skies, but in fact he only gave the idea his valuable and entire support. The sphere of the stars was his eighth heaven. The less elevated heavens, in which he also believed, were invented to explain as well as they might, the proper motions of the sun, moon, and planets.

The philosopher of Stagira said that the motion of his eighth or outermost solid sky was uniform, nor ever troubled by any perturbation. "Within the universe there is," he says, "a fixed and immovable centre, the earth ; and without there is a bounding surface enclosing it on all sides. The outermost part of the universe is the sky. It is filled with heavenly bodies which we know as stars, and it has a perpetual motion, carrying round with it these immortal bodies in its unaltering and unending revolution."

Euclid, to whom we may assign a date of about 275 before our present era, also considered the stars to be set in a solid sphere, having the eye of the observer as centre ; though for him this conception was simply a deduction from exact and fundamental observations, namely, that their revolution took place as a whole, the shape and size of the constellation being never altered.

Cicero, in the last century before Christ, declared himself a believer in the solidity of the sky. According to him the ether was too rarefied to enable it to move the stars, which must therefore require to be fixed to a sphere of their own, independent of the ether.

In the time of Seneca there seem to have been difficulties already raised about the solidity of the heavens, for he only mentions it in the form of a question—" Is the sky solid and of a firm and compact substance ? " (*Questions*, Book ii.)

In the fifth century the idea of the star sphere still lingered, and in the eyes of Simplicius, the commentator of Aristotle, it was not merely an artifice suitable for the representation of the apparent motions, but a firm and solid reality ; while Mahomet and most of the Fathers of the Christian Church had the same conception of these concentric spheres.

It appears then from this review that the phrases " starry vault," and especially " fixed stars," have been used in two very distinct senses. When we meet with them in Aristotle or Ptolemy, it is obvious that they have reference to the crystal sphere of Anaximenas, to which they were supposed to be affixed, and to move with it ; but that later the word " fixed " carried with it the sense of immovable, and the stars were conceived as fixed in this sense, independently of the sphere to which they were originally thought to be attached. Thus Seneca speaks of them as the *fixum et immobilem populum.*

If we would inquire a little further into the supposed nature of this solid sphere, we find that Empedocles considered it to be a solid mass, formed of a portion of the ether which the elementary fire has converted into crystal, and his ideas of the connection between cold and solidification being not very precise, he described it by names that give the best idea of transparence, and, like Lactantius, called it *ritreum cælum,* or said *cælum ærem glaciatum esse,* though we cannot suppose that he made any allusion to what we now call glass, but simply meant some body eminently transparent into which the fire had transformed the air; while so far from having any idea of cold, as we might imagine possible from observations of the snowy tops of mountains, they actually believed in a warm region above the lower atmosphere. Thus Aristotle considers that the spheres heat by their motion the air below them, without being heated themselves, and that there is thus a production of heat. "The motion of the sphere of fixed stars," he says, "is the most rapid, as it moves in a circle with all the bodies attached to it, and the spaces immediately below are strongly heated by the motion, and the heat, thus engendered, is propagated downwards to the earth." This however, strangely enough, does not appear to have prevented their supposing an eternal cold to reign in the regions next below, for Macrobius, in his commentary on Cicero, speaks of the decrease of temperature with the height, and concludes that the extreme zones of the heavens where Saturn moves must be eternally cold; but this they reckoned as part of the atmosphere, beyond whose limits alone was to be found the fiery ether.

It is to the Fathers of the Church that we owe the transmission during the middle ages of the idea of a crystal vault. They conceived a heaven of glass composed of eight or ten superposed layers, something like so many skins in an onion. This idea seems to have lingered on in certain cloisters of southern Europe even into the nineteenth century, for a

venerable Prince of the Church told Humboldt in 1815, that
a large aerolite lately fallen, which was covered with a
vitrified crust, must be a fragment of the crystalline sky.   On
these various spheres, one enveloping without touching
another, they supposed the several planets to be fixed, as we
shall see in a subsequent chapter.

Whether the greater minds of antiquity, such at Plato,
Plutarch, Eudoxus, Aristotle, Apollonius, believed in the
reality of these concentric spheres to carry the planets, or
whether this conception was not rather with them an
imaginary one, serving only to simplify calculation and
assist the mind in the solution of the difficult problem of
their motion, is a point on which even Humboldt cannot
decide.   It is certain, however, that in the middle of the
sixteenth century, when the theory involved no less than
seventy-seven concentric spheres, and later, when the adver-
saries of Copernicus brought them all into prominence to
defend the system of Ptolemy, the belief in the existence of these solid spheres, circles and epicycles, which was
under the especial patronage of the Church, was very
widespread.

Tycho Brahe expressly boasts of having been the first, by
considerations concerning the orbits of the comets, to have
demonstrated the impossibility of solid spheres, and to have
upset this ingenious scaffolding.   He supposed the spaces
of our system to be filled with air, and that this medium,

disturbed by the motion of the heavenly bodies, opposed a resistance which gave rise to the harmonic sounds.

It should be added also that the Grecian philosophers, though little fond of observation, but rejoicing rather in framing systems for the explanation of phenomena of which they possessed but the faintest glimpse, have left us some ideas about the nature of shooting stars and aerolites that come very close to those that are now accepted. "Some philosophers think," says Plutarch in his life of Lysander, "that shooting stars are not detached particles of ether which are extinguished by the atmosphere soon after being ignited, nor do they arise from the combustion of the rarefied air in the upper regions, but that they are rather heavenly bodies which fall, that is to say, which escaping in some way from the general force of rotation are precipitated in an irregular manner, sometimes on inhabited portions of the earth, but sometimes also in the ocean, where of course they cannot be found." Diogenes of Apollonius expresses himself still more clearly: "Amongst the stars that are visible move others that are invisible, to which in consequence we are unable to give any name. These latter often fall to the earth and take fire like that star-stone which fell all on fire near Ægos Potamos." These ideas were no doubt borrowed from some more ancient source, as he believed that all the stars were made of something like pumice-stone. Anaxagoras, in fact, thought that all the heavenly bodies were fragments of rocks which the ether, by the force of its circular motion, had detached from the earth, set fire to, and turned into stars." Thus the Ionic school, with Diogenes of Apollonius, placed the aërolites and the stars in one class, and assigned to all of them a terrestrial origin, though in this sense only, that the earth, being the central body, had furnished the matter for all those that surround it.

Plutarch speaks thus of this curious combination :— " Anaxagoras teaches that the ambient ether is of an igneous nature, and by the force of its gyratory motion it tears off blocks of stone, renders them incandescent, and transforms them into stars." It appears that he explained also by an analogous effect of the circular motion the descent of the Nemæan Lion, which, according to an old tradition, fell out of the moon upon the Peloponnesus. According to Bœckh, this ancient myth of the Nemæan Lion had an astronomical origin, and was symbolically connected in chronology with the cycle of intercalation of the lunar year, with the worship of the moon in Nemæa, and the games by which it was accompanied.

Anaxagoras explains the apparent motion of the celestial sphere from east to west by the hypothesis of a general revolution, the interruption of which, as we have just seen, caused the fall of meteoric stones. This hypothesis is the point of departure of the theory of vortices, which more than two thousand years later, by the labours of Descartes,

L

PLATE VI. —THE NEMEAN LION.

Huyghens, and Hooke, took so prominent a place among the theories of the world.

It may be worth adding with regard to the famous aërolite of Ægos Potamus, alluded to above, that when the heavens were no longer believed to be solid, the faith in the celestial origin of this, as of other aërolites, was for a long time destroyed. Thus Bailly the astronomer, alluding to it, says, "if the fact be true, this stone must have been thrown out by a volcano." Indeed it is only within the last century that it has been finally accepted for fact that stones do fall from the sky. Laplace thought it probable that they came from the moon; but it has now been demonstrated that aërolites, meteors, and shooting stars belong all to one class of heavenly bodies, that they are fragments scattered through space, and circulate like the planets round the sun. When the earth in its motion crosses this heavenly host, those which come near enough to touch its atmosphere leave a luminous train behind them by their heating by friction with the air: these are the *shooting stars*. Sometimes they come so close as to appear larger than the moon, then they are *meteors ;* and sometimes too the attraction of the earth makes them fall to it, and these become *aërolites*.

But to return to our ancient astronomers :—

They believed the heavens to be in motion, not only because they saw the motion with their eyes, but because they believed them to be animated, and regarded motion as the essence of life. They judged of the rapidity of the stars' motion by a very ingenious means. They perceived that it was greater than that of a horse, a bird, an arrow, or even of the voice, and Cleomenas endeavoured to estimate it in the following way. He remarks that when the king of Persia made war upon Greece he placed men at certain intervals, so as to be in hearing of each other, and thus passed on the news from Athens to Susa. Now this news took two days and nights to pass over this distance. The voice therefore only accomplished a fraction of the distance that the stars had accomplished twice in the same time.

The heavens, as we have seen, were not supposed to consist of a single sphere, but of several concentric ones, the arrangement and names of which we must now inquire into.

The early Chaldeans established three. The first was the empyreal heaven, which was the most remote. This, which they called also the solid firmament, was made of fire, but of fire of so rare and penetrating a nature, that it easily passed through the other heavens, and became universally diffused, and in this way reached the earth. The second was the ethereal heaven, containing the stars, which were simply formed of the more compact and denser parts of this substance; and the third heaven was that of the planets. The Persians, however, gave a separate heaven to the sun, and another to the moon.

The system which has enjoyed the longest and most widely-

spread reign is that which places above, or rather round, the
solid firmament a heaven of water—(the nature of which
is not accurately defined), and round this a *primum mobile*,
prime mover, or originator of all the motions, and round all
this the empyreal heaven, or abode of the blessed.   In the most
anciently printed scientific encyclopædia known, the *Margarita
philosophica*, edited in the fifteenth century, that is, two centuries before the adoption of the true system of the world, we
have the curious figure represented on the next page, in which
we find no less than eleven different heavens.   We here see
on the exterior the solid empyreal heaven, which is stated in
the body of the work to be the abode of the blessed and to
be immovable, while the next heaven gives motion , all
within, and is followed by the aqueous heaven, then the crystal
firmament, and lastly by the several heavens of the planets,
sun, and moon.   The revolution of these spheres was not
supposed to take place, like the motion of the earth in modern
astronomy, round an imaginary axis, but round one which had
a material existence, which was provided with pivots moving
in fixed sockets.   Thus Vitruvius, architect to Augustus,
teaches it expressly in these words :—

"The heaven turns continually round the earth and sea
upon an axis, where two extremities are like two pivots that
sustain it : for there are two places in which the Governor of
Nature has fashioned and set these pivots as two centres ; one
is above the earth among the northern stars ; the other is at

the opposite end beneath the earth to the south; and around
these pivots, as round two centres, he has placed little naves,

FIG. 13.

like those of a wheel upon which the heaven turns continually."

Similarly curious ideas we shall find to have prevailel
with respect to the meaning of everything that they observel
in the heavens : thus what a number of opinions have been
hazarded on the nature of the " Milky Way " alone! some of
which we may learn from Plutarch.   The Milky Way, he says,
is a nebulous circle, which constantly appears in the sky, and
which owes its name to its white appearance.   Certain
Pythagoreans assert that when Phaeton lit up the universe,
one star, which escaped from its proper place, set light to the
whole space it passed over in its circular course, and so formed
the Milky Way.   Others thought that this circle was where
the sun had been moving at the beginning of the world.
According to others it is but an optical phenomenon produced
by the reflection of the sun's rays from the vault of the sky
as from a mirror, and comparable with the effects seen in the
rainbow and illuminated clouds.   Metrodorus says it is the
mark of the sun's passage which moves along this circle.
Parmenidas pretends that the milky colour arises from a
mixture of dense and rare air.   Anaxagoras thinks it an
effect of the earth's shadow projected on this part of the
heavens, when the sun is below.   Democritus says that it is
the lustre of several little stars which are very near together,
and which reciprocally illuminate each other.   Aristotle
believes it to be a vast mass of arid vapours, which takes
fire from a glowing tress, above the region of the ether, and
far below that of the planets.   Posidonius says that the

circle is a compound of fire less dense than that of the stars,
but more luminous. All such opinions, except that of
Democritus, are of little value, because founded on nothing ;
perhaps the worst is that of Theophrastus, who said it was
the junction between the two hemispheres, which together
formed the vault of heaven : and that it was so badly made
that it let through some of the light that he supposed to
exist everywhere behind the solid sky.

We now know that the Milky Way, like many of the
nebulæ, is an immense agglomeration of suns. The Milky
Way is itself a nebula, a mass of sidereal systems, with our
own among them, since our sun is a single star in this vast
archipelago of eighteen million orbs. The Greeks called it
the Galaxy. The Chinese and Arabians call it the River of
Heaven. It is the Path of Souls among the North American
Indians, and the Road of S. Jacques de Compostelle among
French peasants.

In tracing the history of ideas concerning the structure
of the heavens among the Greek philosophers, we meet with
other modifications which it will be interesting to recount.
Thus Eudoxus, who paid greater attention than others to
the variations of the motions of the planets, gave more than
one sphere to each of them to represent these observed
changes. Each planet, according to him, has a separate part
of the heaven to itself, which is composed of several
concentric spheres, whose movements, modifying each other,

produce that of the planet. He gave three spheres to the
sun : one which turned from east to west in twenty-four
hours, to represent the diurnal rotation; a second, which
turned about the pole of the ecliptic in 365¼ days, and
produced its annual movement; and a third was added to
account for a certain supposed motion, by which the sun
was drawn out of the ecliptic, and turned about an axis,
making such an angle with that of the ecliptic, as
represented the supposed aberration. The moon also had
three spheres to produce its motions in longitude and latitude,
and its diurnal motion. Each of the other planets had four,
the extra one being added to account for their stations and
retrogressions. It should be added that these concentric
spheres were supposed to fit each other, so that the different
planets were only separated by the thicknesses of these
crystal zones.

Polemarch, the disciple of Eudoxus, who went to Athens
with his pupil Calippus for the express purpose of consulting
Aristotle on these subjects, was not satisfied with the exact-
ness with which these spheres represented the planetary
motions, and made changes in the direction of still greater
complication. Instead of the twenty-six spheres which
represented Eudoxus' system, Calippus established thirty-
three, and by adding also intermediary spheres to prevent
the motion of one planet interfering with that of the adjacent
ones, the number was increased to fifty-six.

There is extant a small work, ascribed to Aristotle, entitled "Letter of Aristotle to Alexander on the system of the world," which gives so clear an account of the ideas entertained in his epoch that we shall venture to give a somewhat long extract from it. The work, it should be said, is not by all considered genuine, but is ascribed by some to Nicolas of Damas, by others to Anaximenas of Lampsacus, a contemporary of Alexander's, and by others to the Stoic Posidonius. It is certain, however, that Aristotle paid some attention to astronomy, for he records the rare phenomena of an eclipse of Mars by the moon, and the occultation of one of the Gemini by the planet Jupiter, and the work may well be genuine. It contains the following:—

"There is a fixed and immovable centre to the universe. This is occupied by the earth, the fruitful mother, the common focus of every kind of living thing. Immediately surrounding it on all sides is the air. Above this in the highest region is the dwelling-place of the gods, which is called the heavens. The heavens and the universe being spherical and in continual motion, there must be two points on opposite sides, as in a globe which turns about an axis, and these points must be immovable, and have the sphere between them, since the universe turns about them. They are called the poles. If a line be drawn from one of these points to the other it will be the diameter of the universe, having the earth in the centre and the two poles at the extremities; of these two poles the northern one is always visible above our horizon, and is called the Arctic pole; the other, to the south, is always invisible to us—it is called the Antarctic pole.

"The substance of the heavens and of the stars is called ether; not that it is composed of flame, as pretended by some who have not considered its nature, which is very different from that of fire, but it is so called because it has an eternal circular motion, being a divine and incorruptible element, altogether different from the other four.

" Of the stars contained in the heavens some are fixed, and turn with the heavens, constantly maintaining their relative positions. In their middle portion is the circle called the *zoophore*, which stretches obliquely from one tropic to the other, and is divided into twelve parts, which are the twelve signs (of the zodiac). The others are wandering stars, and move neither with the same velocity as the fixed stars, nor with a uniform velocity among themselves, but all in different circles, and with velocities depending on the distances of these circles from the earth.

"Although all the fixed stars move on the same surface of the heavens, their number cannot be determined. Of the movable stars there are seven, which circulate in as many concentric circles, so arranged that the lower circle is smaller than the higher, and that the seven so placed one within the other are all within the spheres of the fixed stars.

" On the nearer, that is inner, side of this ethereal, immovable, unalterable, impassible nature is placed our movable, corruptible, and mortal nature. Of this there are several kinds, the first of which is fire, a subtle inflammable essence, which is kindled by the great pressure and rapid motion of the ether. It is in this region of air, when any disturbance takes place in it, that we see kindled shooting-stars, streaks of light, and shining motes, and it is there that comets are lighted and extinguished.

" Below the fire comes the air, by nature cold and dark, but which is warmed and enflamed, and becomes luminous by its motion. It is in the region of the air, which is passive and changeable in any manner, that the clouds condense, and rain, snow, frost, and hail are formed and fall to the earth. It is the abode of stormy winds, of whirlwinds, thunder, lightning, and many other phenomena.

" The cause of the heaven's motion is God. He is not in the centre, where the earth is a region of agitation and trouble, but he is above the outermost circumference, which is the purest of all regions, a place which we call rightly *ouranos*, because it is the highest part of the universe, and *olympos*, that is, perfectly bright, because it is altogether separated from everything like the shadow and disordered movements which occur in the lower regions."

We notice in this extract a curious etymology of the word ether, namely, as signifying perpetual motion (ἀεὶ τεεῖν), though it is more probable that its true, as its more generally accepted derivation is from αἴθειν, to burn or shine, a meaning doubtless alluded to in a remarkable passage of Hippocrates, Περὶ Σάρκων. " It appears to me," he says, " that what we call the principle of heat is immortal, that it knows all, sees all, hears all, perceives all, both in the past and in the future. At the time when all was in confusion, the greater part of this principle rose to the circumference of the universe; it is this that the ancients have called *ether*."

The first Greek that can be called an astronomer was Thales, born at Miletus 641 B.C., who introduced into Greece the elements of astronomy. His opinions were these: that the stars were of the same substance as the earth, but that they were on fire; that the moon borrowed its light from the sun, and caused the eclipses of the latter, while it was itself eclipsed when it entered the earth's shadow: that the earth was round, and divisible into five zones, by means of five circles, *i.e.* the Arctic and Antartic, the two tropics, and the equator; that the latter circle is cut obliquely by the ecliptic, and perpendicularly by the meridian. Up to his time no division of the sphere had been made beyond the description of the constellations. These opinions do not appear to have been rapidly spread, since Herodotus, one of the finest intellects of Greece, who lived two centuries later, was still so ill-instructed as to say, in speaking of an eclipse, " The sun abandoned its place, and night took the place of day."

Anaxagoras, of whom we have spoken before, asserted that the sun was a mass of fire larger than the Peloponnesus. Plutarch says he regarded it as a burning stone, and Diogenes Laertius looked upon it as hot iron. For this bold idea he was persecuted. They considered it a crime that he taught the causes of the eclipses of the moon, and pretended that the sun is larger than it looks. He first taught the existence of one God, and he was taxed with impiety and treason against his country. When he was condemned to death, "Nature," he said, "has long ago condemned me to the same; and as to my children, when I gave them birth I had no doubt but they would have to die some day." His disciple Pericles, however, defended him so eloquently that his life was spared, and he was sent into exile.

Pythagoras, who belonged to the school of Thales, and who travelled in Phœnicia, Chaldea, Judæa, and Egypt, to learn their ideas, ventured, in spite of the warnings of the priests, to submit to the rites of initiation at Heliopolis, and thence returned to Samos, but meeting with poor reception there, he went to Italy to teach. From him arose the *Italian School*, and his disciples took the name of philosophers (lovers of wisdom) instead of that of sages. We shall learn more about him in the chapter on the Harmony of the Spheres.

His first disciple, Empedocles, famous for the curiosity which led him to his death in the crater of Ætna, as the story goes, thought that the true sun, the fire that is in the centre of the universe, illuminated the other hemisphere, and that what we see is only the reflected image of that, which is invisible to us, and all of whose movements it follows.

His disciple, Philolaus, also taught that the sun was a mass of glass, which sent us by reflection all the light that it scattered through the universe. We must not, however, forget that these opinions are recorded by historians who probably did not understand them, and who took in the letter what was only intended for a comparison or figure.

If we are to believe Plutarch, Xenophanes, who flourished about 360 B.C., was very wild in his opinions. He thought the stars were lighted every night and extinguished every morning; that the sun is a fiery cloud; that eclipses take place by the sun being extinguished and afterwards re-kindled; that the moon is inhabited, but is eighteen times larger than the earth; that there are several suns and several moons for giving light to different countries. This can only be matched by those who said the sun went every night through a hole in the earth round again to the east; or that it went above ground, and if we did not see it going back it was because it accomplished the journey in the night.

Parmenidas was the disciple of Xenophanes. He divided the earth, like Thales, into zones; and he added that it was suspended in the centre of the universe, and that it did not fall because there was no reason why it should move in one direction rather than another. This argument is perfectly

philosophical, and illustrates a principle employed since the
time of Archimedes, and of which Leibnitz made so much
use.

Such are some of the general ideas which were held by
the Greeks and others on the nature of the heavens, omitting
that of Ptolemy, of which we shall give a fuller account
hereafter. We see that they were all affected by the domi-
nant idea of the superiority of the earth over the rest of the
universe, and were spoiled for want of the grand conception
of the immensity of space. The universe was for them a
closed space, outside of which there was *nothing*; and they
busied themselves with metaphysical questions as to the
possibility of space being infinite. In the meantime their
conceptions of the distances separating us from other visible
parts of the universe were excessively cramped. Hesiod, for
instance, thinks to give a grand idea of the size of the
universe by saying that Vulcan's anvil took seven days to
fall from heaven to earth, when in reality, as now calculated,
it would take no less than seventy-two years for the light,
even travelling at a far greater rate, to reach us from one
of the nearest of the fixed stars.

CHAPTER VII.

NATURE presents herself to us under various aspects. At times, it may be, she presents to us the appearance of discord, and we fail to perceive the unity that pervades the whole of her actions. At others, however, and most often to an instructed mind, there is a concord between her various powers, a harmony even in her sounds, that will not escape us, Even the wild notes of the tempest and the bass roll of the thunder form themselves into part of the grand chorus which in the great opera are succeeded by the solos of the evening breeze, the songs of birds, or the ripple of the waves. These are ideas that would most naturally present themselves to contemplative minds, and such must have been the students of the silent, but to them harmonious and tuneful, star-lit sky, under the clear atmosphere of Greece. The various motions they observed became indissolubly connected in their minds with music, and they did not doubt that the heavenly spheres made harmony, if imperceptible to human

M

ears. But their ideas were more precise than this. They
discovered that harmony depended on number, and they
attempted to prove that whether the music they might make
were audible or not, the celestial spheres had motions which
were connected together in the same way as the numbers
belonging to a harmony. The study of their opinions on this
point reveals some very curious as well as very interesting ideas. We may commence by referring to an ancient
treatise by Timæus of Locris on the soul of the universe.
To him we owe the first serious exposition of the complete
harmonic cosmography of Pythagoras. We must premise
that, according to this school, God employed all existing
matter in the formation of the universe—so that it compre-
hends all things, and all is in it. "It is a unique, perfect,
and spherical production, since the sphere is the most perfect
of figures; animated and endowed with reason, since that
which is animated and endowed with reason is better than
that which is not."

So begins Timæus, and then follows, as a quotation from
Plato, a comparison of the earth to what would appear to
us nowadays to be a very singular animal. Not only, says
Plato, is the earth a sphere, but this sphere is perfect, and
its maker took care that its surface should be perfectly
uniform for many reasons. The universe in fact has no need
of eyes, since there is nothing outside of it to see; nor yet
of ears, since there is nothing but what is part of itself to

make a sound ; nor of breathing organs, as it is not surrounded by air: any organ that should serve to take in
nourishment, or to reject the grosser parts, would be
absolutely useless, for there being nothing outside it, it
could not receive or reject anything. For the same reason
it needs no hands with which to defend itself, nor yet of feet
with which to walk. Of the seven kinds of motion, its
author has given it that which is most suitable for its figure
in making it turn about its axis, and since for the execution
of this rotatory motion no arms or legs are wanted, its maker
gave it none.

With regard to the soul of the universe, Plato, according to
Timæus, says that God composed it " of a mixture of the divisible
and indivisible essences, so that the two together might be
united into one, uniting two forces, the principles of two kinds
of motion, one that which is *always the same*, and the other
that which is *always changing*. The mixture of these two
essences was difficult, and was not accomplished without
considerable skill and pains. The proportions of the
mixture were according to harmonic numbers, so chosen
that it is possible to know of what, and by what rule,
the soul of the universe is compounded."

By harmonic numbers Timæus means those that are
proportional to those representing the consonances of the
musical scale. The consonances known to the ancients were
three in number : the diapason, or octave, in the proportion of

2 to 1, the diapent, or fifth, in that of 3 to 2, and the diatess-saron, or fourth, in that of 4 to 3; when to these are joined the tones which fill the intervals of the consonances, and are in the proportion of 9 to 8, and the semitones in that of 256 to 243, all the degrees of the musical scale is complete.

The discovery of these harmonic numbers is due to Pythagoras. It is stated that when passing one day near a forge, he noticed that the hammers gave out very accurate musical concords. He had them weighed, and found that of those which sounded the octave, one weighed twice as much as the other; that of those which made a perfect fifth, one weighed one third more than the other, and in the case of a fourth, one quarter more. After having tried the hammers, he took a musical string stretched with weights, and found that when he had applied a given weight in the first instance to make any particular note, he had to double the weight to obtain the octave, to add one third extra only to obtain a fifth, a quarter for the fourth, and eight for one tone, and about an eighteenth for a half-tone ; or more simply still, he stretched a cord once for all, and then when the whole length sounded any note, when stopped in the middle it gave the octave, at the third it gave the fifth, at the quarter the fourth, at the eighth the tone, and at the eighteenth the semi-tone.

Since the ancients conceived of the soul by means of motion, the quantity of motion developed in anything was their measure of the quantity of its soul. Now the motion of the heavenly bodies seemed to them to depend on their distance from the centre of the universe, the fastest being those at the circumference of the whole. To determine the relative degrees of velocity, they imagined a straight line drawn outwards from the centre of the earth, as far as the empyreal heaven, and divided it according to the proportions of the musical scale, and these divisions they called the harmonic degrees of the soul of the universe. Taking the earth's radius for the first number, and calling it unity, or, in order to avoid fractions, denoting it by 384, the second degree, which is at the distance of an harmonic third, will be represented by 384 plus its eighth part, or 432. The third degree will be 432, plus its eighth part, or 486. The fourth, being a semitone, will be as 243 to 256, which will give 512; and so on. The eighth degree will in this way be the double of 384 or 768, and represents the first octave.

They continued this series to 36 degrees, as in the following table :—

The Earth.

Mi	$384 + \frac{1}{8} = 432$
Re	$432 + \frac{1}{8} = 486$
Ut	$486 : 512 :: 243 : 256$
Si	$512 + \frac{1}{8} = 576$
La	$576 + \frac{1}{8} = 648$
Sol	$648 + \frac{1}{8} = 729$
Fa	$729 : 768 :: 243 : 256$
Mi	$768 + \frac{1}{8} = 864$
Re	$864 + \frac{1}{8} = 972$
Ut	$972 : 1024 :: 243 : 256$

Si . .	. . . $1024 + \frac{1}{8} = 1152$
La . .	. . . $1152 + \frac{1}{8} = 1296$
Sol .	. . . $1296 + \frac{1}{8} = 1458$
Fa .	. . . $1458 : 1536 : : 243 : 256$
Mi . .	. . . $1536 + \frac{1}{8} = 1728$
Re	. . . $1728 + \frac{1}{8} = 1944$
Ut .	. . . . $1944 : 2048 : 243 : 256$
Si . .	. . . $2048 + 139 = 2187$
Si 2 .	. . . $2187 : 2304 : : 243 : 256$
La .	. . . $2304 + \frac{1}{8} = 2592$
Sol .	. . . $2592 + \frac{1}{8} = 2916$
Fa .	. . . $2916 : 3072 : : 243 : 256$
Mi .	. . . $3072 + \frac{1}{8} = 3456$
Re	. . . $3457 + \frac{1}{8} = 3888$
Ut . . .	. . . $3888 + \frac{1}{8} = 4374$
Si .	. . . . $4374 : 4608 : : 243 : 256$
La .	. . . $4608 + \frac{1}{8} = 5184$
Sol . .	. . . $5184 + \frac{1}{8} = 5832$
Fa .	. . $5832 : 6144 : : 243 : 256$
Mi . .	. $6144 + 417 = 6561$
Mi ♭ .	. . . $6561 : 6912 : : 243 : 256$
Re .	. . $6912 + \frac{1}{8} = 7776$
Ut .	. . $7776 + \frac{1}{8} = 8748$
Si	. . . $8748 : 9216 : : 243 : 256$
La .	. . . $9216 + \frac{1}{8} = 10368$
Sol	. . . $10368 = 384 + 27$

The empyreal heaven.
Sum of all the terms, 114,695.

This series they considered a complete one, because by
taking the terms in their proper intervals, the last becomes
27 times the original number, and in the school of Pythagoras
this 27 had a mystic signification, and was considered as the
perfect number.

The reason for considering 27 a perfect number was curious.

It is the sum of the first linear, square, and cubic numbers added to unity. First there is 1, which represents the point, then 2 and 3, the first linear numbers, even and uneven, then 4 and 9, the first square or surface numbers, even and uneven, and the last 8 and 27, the first solid or cubic numbers, even and uneven, and 27 is the sum of all the former. Whence, taking the number 27 as the symbol of the universe, and the numbers which compose it as the elements, it appeared right that the soul of the universe should be composed of the same elements.

On this scale of distances, with corresponding velocities, they arranged the various planets, and the universe comprehended all these spheres, from that of the fixed stars (which was excluded) to the centre of the earth. The sphere of the fixed stars was the common envelope, or circumference of the universe, and Saturn, immediately below it, corresponded to the thirty-sixth tone, and the earth to the first, and the other planets . with the sun and moon at the various harmonic distances.

They reckoned one tone from the earth to the moon, half a tone from the moon to Mercury, another half-tone to Venus, one tone and a half from Venus to the sun, one from the sun to Mars, a semitone from Mars to Jupiter, half a tone from Jupiter to Saturn, and a tone and a half from Saturn to the fixed stars; but these distances were not, as we shall see, universally agreed upon.

According to Timæus, the sphere of the fixed stars, which

contains within it no principle of contrariety, being entirely divine and pure, always moves with an equal motion in the same direction from east to west. But the stars which are within it, being animated by the mixed principle, whose composition has been just explained, and thus containing two contrary forces, yield on account of one of these forces to the motion of the sphere of fixed stars from east to west, and by the other they resist it, and move in a contrary direction, in proportion to the degree with which they are endowed with each; that is to say, that the greater the proportion of the material to the divine force that they possess, the greater is their motion from west to east, and the sooner they accomplish their periodic course. Now the amount of this force depends on the matter they contain. Thus, according to this system, the planets turn each day by the common motion with all the heavens about the earth from east to west, but they also retrograde towards the east, and accomplish their periods according to their component parts.

The additions which Plato made to this theory have always been a proverb of obscurity, and none of his commentators have been able to make anything of them, and very possibly they were never intended to.

So far the harmony of the heavenly bodies has been explained with reference to numbers only, and we may add to this that they reckoned 126,000 stadia, or 14,286 miles, to represent a tone, which was thus the distance of the earth to

the moon, and the same measurement made it 500,000 from
the earth to the sun, and the same distance from the sun
to the fixed stars.

But Plato teaches in his *Republic* that there is actual
musical harmony between the planets. Each of the spheres,
he said, carried with it a Siren, and each of these sounding a
different note, they formed by their union a perfect concert,
and being themselves delighted with their own harmony,
they sang divine songs, and accompanied them by a sacred
dance. The ancients said there were nine Muses, eight
of whom, according to Plato, presided over celestial, and
the ninth over terrestrial things, to protect them from disorder
and irregularity.

Cicero and Macrobius also express opinions on this
harmonious concert. Such great motions, says Cicero,
cannot take place in silence, and it is natural that the
two extremes should have related sounds as in the octave.
The fixed stars must execute the upper note, and the moon
the base. Kepler has improved on this, and says Jupiter
and Saturn sing bass, Mars takes the tenor, the earth and
Venus are contralto, and Mercury is soprano! True, no one
has ever heard these sounds, but Pythagoras himself may
answer this objection. We are always surrounded, he says,
by this melody, and our ears are accustomed to it from our
birth, so that, having nothing different to compare it with, we
cannot perceive it.

We may here recall the further development of the idea
of the soul of the universe, which was the source of this
harmony, and endeavour to find a rational interpretation of
their meaning.   They said that nature had made the animals
mortal and ephemeral, and had infused their souls into them,
as they had been extracts from the sun or moon, or even from
one of the planets.   A portion of the unchangeable essence
was added to the reasoning part of man, to form a germ of
wisdom in privileged individuals.   For the human soul there
is one part which possesses intelligence and reason, and
another part which has neither the one nor the other.

The various portions of the general soul of the universe
resided, according to Timæus, in the different planets, and
depended on their various characters.   Some portions were
in the moon, others in Mercury, Venus, or Mars, and so on,
and thus they give rise to the various characters and
dispositions that are seen among men.   But to these parts
of the human soul that are taken from the planets is joined
a spark of the supreme Divinity, which is above them all,
and this makes man a more holy animal than all the rest,
and enables him to have immediate converse with the
Deity himself.   All the different substances in nature were
supposed to be endowed with more or less of this soul, according to their material nature or subtilty, and were placed in
the same order along the line, from the centre to the circumference, on which the planets were situated, as we have seen

above. In the centre was the earth, the heaviest and grossest of all, which had but little if any soul at all. Between the earth and the moon, Timæus placed first water, then the air, and lastly elementary fire, which he considered to be principles, which were less material in proportion as they were more remote and partook of a larger quantity of the soul of the universe. Beyond the moon came all the planets, and thus were filled up the greater number of the harmonic degrees, the motions of the various bodies being guided by the principle enunciated above.

When we carefully consider this theory we find that by a slight change of name we may bring it more into harmony with modern ideas. It would appear indeed that the ancients called that " soul " which we now call " force," and while we say that this force of attraction is in proportion to the masses and the inverse square of the distance, they put it that it was proportional to the matter, and to the divine substance on which the distance depended. So that we may interpret Timæus as stating this proposition : *The distances of the stars and their forces are proportional among themselves to their periodic times.* " Some people," says Plutarch, " seek the proportions of the soul of the universe in the velocities (or periodic times), others in the distances from the centre; some in the masses of the heavenly bodies, and others more acute in the ratios of the diameters of their orbits. It is probable that the mass of each planet, the

intervals between the spheres and the velocities of their
motions, are like well-tuned musical instruments, all proportional harmonically with each other and with all other parts
of the universe, and by necessary consequence that there are
the same relative proportions in the soul of the universe by
which they were formed by the Deity."

It is marvellous how deeply occupied were all the best
minds in Greece and Italy on this subject, both poets and
philosophers; Ocellus, Democritus, Timæus, Aristotle, and
Lucretius have all left treatises on the same subject, and
almost with the same title, "The Nature of the Universe."

Though somewhat similar to that of Timæus, it will be
interesting to give an account of the ideas of one of these,
Ocellus of Lucania.

Ocellus represents the universe as having a spherical form.
This sphere is divided into concentric layers; above that of
the moon they were called celestial spheres, while below it and
inwards as far as the centre of the earth they were called the
elementary spheres, and the earth was the centre of them all.

In the celestial spheres all the stars were situated, which
were so many gods, and among them the sun, the largest
and most powerful of all. In these spheres is never any
disturbance, storm, or destruction, and consequently no re-
paration, no reproduction, no action of any kind was required
on the part of the gods. Below the moon all is at war, all
is destroyed and reconstructed, and here therefore it is that

generations are possible. But these take place under the influence of the stars, and particularly that of the sun, which in its course acts in different ways on the elementary spheres, and produces continual variations in them, from whence arises the replenishing and diversifying of nature. It is the sun that lights up the region of fire, that dilates the air, melts the water, and renders fertile the earth, in its daily course from east to west, as well as in this annual journey into the two tropics. But to what does the earth owe its germs and its species ? According to some philosophers these germs were celestial ideas which both gods and demons scattered from above over every part of nature, but according to Ocellus they arise continually under the influence of the heavenly bodies. The divisions of the heavens were supposed to separate the portion that is unalterable from that which is in ceaseless change. The line dividing the mortal from the immortal is that described by the moon : all that lies above that, inclusive, is the habitation of the gods ; all that lies below is the abode of nature and discord ; the latter tending constantly to destruction, the former to the reconstruction of all created things.

Ideas such as these, of which we could give other examples more remotely connected with harmony, whatever amount of truth we may discover in them, prove themselves to have been made before the sciences of observation had enabled men to make anything better than empty theories, and to support them with false logic. No better example of the latter can perhaps be mentioned here than the way in which Ocellus pretends to prove that the world is eternal. "The universe," he says, "*having* always existed, it follows that everything in it and every arrangement of it must always have been as it is now. The several parts of the universe *having* always existed with it, we may say the same of the parts of these parts; thus the sun, the moon, the fixed stars, and the planets have always existed with the heavens; animals, vegetables, gold, and silver with the earth; the currents of air, winds, and changes from hot to cold, from cold to hot, with the air. *Therefore* the heaven, with all that it now contains; the earth, with all that it produces and supports; and lastly, the whole aërial region, with all its phenomena, have always existed." When this system of argument passed away, and exact observation took its place, it was soon found that so far from what the ancients had argued *must be* really being the case, no such rela on as they indicated between the distances or velocities of the planets could be traced, and therefore no harmony in the heavens in this sense. It is not indeed that we can say no sounds exist because we hear none; but considering harmony really to consist of the relations of numbers, no such relations exist between the planets' distances, as measured now of course from the sun, instead of being, as then, imagined from the earth.

The gamut is nothing else than the series of numbers:—

do	re	mi	fa	sol	la	si	do
1	$\frac{9}{8}$	$\frac{5}{4}$	$\frac{4}{3}$	$\frac{3}{2}$	$\frac{5}{3}$	$\frac{15}{8}$	2

and is independent of our perception of the corresponding notes. A concert played before a deaf assembly would be a concert still. If one note is made by 10,000 vibrations per second, and another by 20,000, we should hear them as an octave, but if one had only 10 and the other 20, they would still be an octave, though inaudible as notes to us; so too we may speak even of the harmony of luminous vibrations of ether, though they do not affect our ears.

The velocities of the planets do not coincide with the terms of this series. The nearer they are to the sun the faster is their motion, Mercury travelling at the mean rate of 55,000 metres a second, Venus, 36,800, the earth 30,550, Mars 24,448, Jupiter 13,000, Saturn 9,840, Uranus 6,800, and Neptune 5,500, numbers which are in the proportion roundly of 100, 67, 55, 44, 24, 16, 12, 10, which have no sufficient relation to the terms of an harmonic series, to make any harmony obvious.

Returning, however, to the ancient philosophers, we are led by their ideas about the soul of the universe to discover the origin of their gods and natural religion. They were persuaded that only living things could move, and consequently that the moving stars must be endowed with superior intelligence. It may very well be that from the number seven of the planets, including the sun and moon, which were their earliest gods, arose the respect and superstition with which all nations, and especially the Orientals, regarded that number. From these arose the seven superior angels that are found in the theologies of the Chaldeans, Persians, and Arabians; the seven gates of Mithra, through which all souls must pass to reach the abode of bliss; the seven worlds of purification of the Indians, and all the other applications of the number seven which so largely figure in Judaism, and have descended from it to our own time. On the other hand, as we have seen, this number seven may have been derived from the number of the stars in the Pleiades.

We have noticed in our chapter on the History of the Zodiac how the various signs as they came round and were thought to influence the weather and other natural phenomena, came at last to be worshipped. Not less, of course, were the sun and moon deified, and that by nations who had no zodiac. Among the Egyptians the sun was painted in different forms according to the time of year, very much as he is represented in our own days in pictures of the old and new years. At the winter solstice with them he was an infant, at the spring equinox he was a young man in summer a man in full age with flowing beard, and in the autumn an old man. Their fable of Osiris was founded on the same idea. They represented the sun by the hawk, and the moon by the Ibis, and to these two, worshipped under the names of Osiris and Isis they attributed the government of the world, and built a city, Heliopolis, to the former, in the temple of which they placed his statue.

The Phenicians in the same way, who were much influenced by ideas of religion, attributed divinity to the sun, moon, and stars, and regarded them as the sole causes of the production and destruction of all things. The sun, under the name of Hercules, was their great divinity.

The Ethiopians worshipped the same, and erected the famous table of the sun. Those who lived above Meroë, admitted the existence of eternal and incorruptible gods, among which they included the sun, moon, and the universe. Like the Incas of Peru, they called themselves the children of the sun, whom they regarded as their common father.

The moon was the great divinity of the Arabs. The Saracens called it Cabar, or the great, and its crescent still adorns the religious monuments of the Turks. Each of their tribes was under the protection of some particular star. Sabeism was the principal religion of the east. The heavens and the stars were its first object.

In reading the sacred books of the ancient Persians contained in the *Zendavesta*, we find on every page invocations addressed to Mithra, to the moon, the stars, the elements, the mountains, the trees, and every part of nature.

N

The ethereal fire circulating through all the universe, and of which the sun is the principal focus, was represented among the fire-worshippers by the sacred and perpetual fire of their priests. Each planet had its own particular temple, where incense was burnt in its honour. These ancient peoples embodied in their religious systems the ideas which, as we have seen, led among the Greeks to the representation of the harmony of heaven. All the world seemed to them animated by a principle of life which circulated through all parts, and which preserved it in an eternal activity. They thought that the universe lived like man and the other animals, or rather that these latter only lived because the universe was essentially alive, and communicated to them for an instant an infinitely small portion of its own immortality. They were not wise, it may be, in this, but they appear to have caught some of the ideas that lie at the basis of religious thought, and to have traced harmony where we have almost lost the perception of it.

CHAPTER VIII.

In our former chapters we have gained some idea of the general structure of the heavens as represented by ancient philosophers, and we no longer require to know what was thought in the infancy of astronomy, when any ideas promulgated were more or less random ones ; but in this chapter we hope to discuss those arrangements of the heavenly bodies which have been promulgated by men as complete systems, and were supposed to represent the totality of the facts.

The earliest thoroughly-established system is that of Ptolemy. It was not indeed invented by him. The main ideas had been entertained long before his time, but he gave it consistence and a name.

We obtain an excellent view of the general nature of this system from Cicero. He writes :—

" The universe is composed of nine circles, or rather of nine moving globes. The outermost sphere is that of the heavens which surrounds all the others, and on which are fixed the stars. Beneath this revolve seven other globes, carried round by a motion in a direction contrary to that of the heavens. On the first circle revolves the star which men call Saturn ; on the second Jupiter shines, that beneficent and propitious star to human eyes ; then follows Mars, ruddy and awful. Below, and occupying the middle region, revolves the Sun, the chief, prince, and moderator of the other stars, the soul of the world, whose immense globe spreads its light through space. After him come, like two companions, Venus and Mercury. Lastly, the lowest globe is occupied by the moon, which borrows its light from the star of day. Below this last celestial circle, there is nothing but what is mortal and corruptible, except the souls given by a beneficent Divinity to the race of men. Above the moon all is eternal. The earth, situated in the centre of the world, and separated from heaven on all sides, forms the ninth sphere ; it remains immovable, and all heavy bodies are drawn to it by their own weight."

The earth, we should add, is surrounded by the sphere of air, and then by that of fire, and by that of ether and the meteors.

With respect to the motions of these spheres. The first circle described about the terrestrial system, namely, that of the moon, was accomplished in 27 days, 7 hours, and 43 minutes. Next to the moon, Mercury in the second, and Venus in the third, and the sun in the fourth circle, all turned about the earth in the same time, 365 days, 5 hours, and 49

minutes. But these planets, in addition to the general movement, which carried them in 24 hours round from east to
west and west to east, and the annual revolution, which made

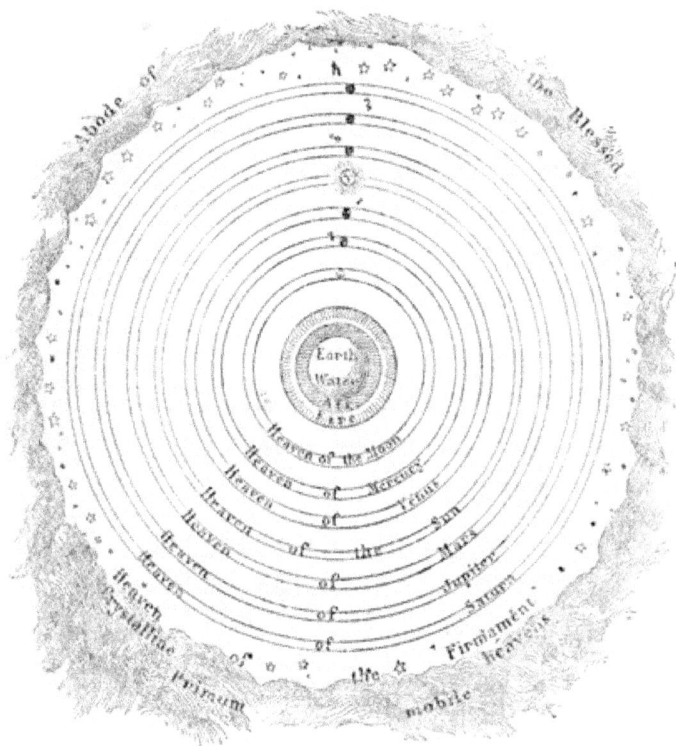

FIG. 14.—PTOLEMY'S ASTRONOMICAL SYSTEM.

them run through the zodiacal circle, had a third motion by
which they described a circle about each point of their orbit
taken as a centre.

The fifth sphere, carrying Mars, accomplished its revolution

in two years.   Jupiter took 11 years, 313 days, and 19 hours
to complete his orbit, and Saturn in the seventh sphere took
29 years and 169 days.   Above all the planets came the
sphere of the fixed stars, or Firmament, turning from east to
west in 24 hours with inconceivable rapidity, and endued
also with a proper motion from west to east, which was
measured by Hipparchus, and which we now call the precession of the equinoxes, and know that it has a period of
25,870 years.   Above all these spheres, a *primum mobile*
gave motion to the whole machine, making it turn from east
to west, but each planet and each fixed star made an effort
against this motion, by means of which each of them accomplished their revolution about the earth in greater or less
time, according to its distance, or the magnitude of the orbit
it had to accomplish.

One immense difficulty attended this system.   The apparent
motions of the planets is not uniform, for sometimes they are
seen to advance from west to east, when their motion is called
direct, sometimes they are seen for several nights in succession at the same point in the heavens, when they are called
stationary, and sometimes they return from east to west, and
then their motion is called *retrograde*.

We know now that this apparent variation in the motion
of the planets is simply due to the annual motion of the
earth in its orbit round the sun.   For example, Saturn describes its vast orbit in about thirty years, and the earth

describes in one year a much smaller one inside. Now if the earth goes faster in the same direction as Saturn, it is plain that Saturn will be left behind and appear to go backwards, while if the earth is going in the same direction the velocity of Saturn will appear to be decreased, but his direction of motion will appear unaltered.

To explain these variations, however, according to his system, Ptolemy supposed that the planets did not move exactly in the circumference of their respective orbits, but about an *ideal centre*, which itself moved along this circumference. Instead therefore of describing a circle, they described parts of a series of small circles, which would combine, as is easy to see, into a series of uninterrupted waves, and these he called *Epicycles*.

Another objection, which even this arrangement did not overcome, was the variation of the size of the planets. To overcome this Hipparchus gave to the sphere of each planet a considerable thickness, and saw that the planet did not turn centrally round the earth, but round a centre of motion placed outside the earth. Its revolution took place in such a manner, that at one time it reached the inner boundary, at another time the outer boundary of its spherical heaven.

But this reply was not satisfactory, for the differences in the apparent sizes proved by the laws of optics such a prodigious difference between their distances from the earth at the times of conjunction and opposition, that it would be extremely difficult to imagine spheres thick enough to allow of it.

It was a gigantic and formidable piece of machinery to which it was necessary to be continually adding fresh pieces to make observation accord with theory. In the thirteenth century, in the times of the King-Astronomer, Alphonso X. of Castile, there were already seventy-five circles, one within the other. It is said that one day he exclaimed, in a full

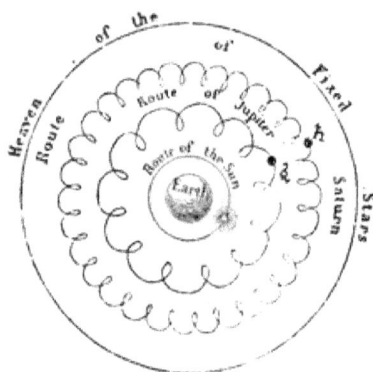

FIG. 15.—THE EPICYCLES OF PTOLEMY.

assemblage of bishops, that if the Deity had done him the honour to ask his advice before creating the world, he could have told Him how to make it a little better, or at all events more simply. He meant to express how unworthy this complication was of the dignity of nature.

Fracastor, in his *Homocentrics*, says that nothing is more monstrous or absurd than all the excentrics and epicycles of

Ptolemy, and proposes to explain the difference of velocity
in the planets at different parts of their orbits by the medium
offering greater or less resistance, and their alteration in
apparent size by the effect of refraction.

The essential element of this system was that it took
appearances for realities, and was founded on the assumption
that the earth is fixed in the centre of the universe, and of
course therefore neglected all the appearances produced
by its motion, or had to explain them by some peculiarity
in the other planets.

Although it was corrected from time to time to make
it accord better with observation, it was the same essentially
that was taught officially everywhere. It reigned supreme
in Egypt, Greece, Italy, and Arabia, and in the great school
of Alexandria, which consolidated it and enriched it by its
own observations.

But though the same in essence, the details, and especially
the means of overcoming the difficulties raised by increased
observations, have much varied, and it will be interesting
and instructive to record some of the chief of them.

One of the most important influences in modifying the
astronomical systems taught to the world has been that
of the Fathers of the Christian Church. When, after five
centuries of patient toil, of hopes, ambitions, and discussions,
the Christian Church took possession of the thrones and
consciences of men, they founded their physical edifice on

the ancient system, which they adapted to their special
wants. With them Aristotle and Ptolemy reigned supreme.
They decreed that the earth constituted the universe, that
the heavens were made for it, that God, the angels, and the
saints inhabited an eternal abode of joy situated above the
azure sphere of the fixed stars, and they embodied this
gratifying illusion in all their illuminated manuscripts, their
calendars, and their church windows.

The doctors of the Church all acknowledged a plurality
of heavens, but they differed as to the number. St. Hilary
of Poitiers would not fix it, and the same doubt held St.
Basil back; but the rest, for the most part borrowing their
ideas from paganism, said there were six or seven, or up to ten.
They considered these heavens to be so many hemispheres
supported on the earth, and gave to each a different name.
In the system of Bede, which had many adherents, they
were the Air, Ether, Fiery Space, Firmament, Heaven of
the Angels, and Heaven of the Trinity.

The two chief varieties in the systems of the middle ages
may be represented as follows:—

Those who wished to have everything as complete as
possible combined the system of Ptolemy with that of the
Fathers of the Church, and placed in the centre of the earth
the infernal regions which they surrounded by a circle.
Another circle marked the earth itself, and after that the
surrounding ocean, marked as water, then the circle of air,

and lastly that of fire.   Enveloping these, and following one after the other, were the seven circles of the seven planets ; the eighth represented the sphere of the fixed stars on the firmament, then came the ninth heaven, then a tenth, the *cœlum cristallinum*, and lastly an eleventh and outermost, which was the empyreal heaven, where dwelt the cherubim and seraphim, and above all the spheres was a throne on which sat the Father, as Jupiter Olympus.

The others who wished for more simplicity, represented the earth in the centre of the universe, with a circle to indicate the ocean, the second sphere was that of the moon ; the third was that of the sun ; on the fourth were placed the four planets, Jupiter, Mars, Venus and Mercury ; there was a fifth for the space outside the planets, and the last outside one was the firmament ; altogether seven spheres instead of eleven. As a specimen of the style of representation of the astronomical systems of the middle ages, we may take the figure on the following page :—

Here we see the earth placed immovable in the centre of the universe, and represented by a disc traversed by the Mediterranean, and surrounded by the ocean.   Round this are circumscribed the celestial spheres.   That of the moon first, then that of Mercury, in which several constellations, as the Lyre, Cassiopeia, the Crown, and others, are roughly indicated, then comes the sphere of Venus with Sagittarius and the Swan.   After this comes the *celestis paradisus,* and the legend that, " the paradise to which Paul

FIG. 16.—HEAVENS OF THE MIDDLE AGES.

was raised is in this third locality; some of these must reach
to us, since in them repose the souls of the prophets." In

the other circles are yet other constellations : for example
Pegasus, Andromeda, the Dog, Argo, the He-goat, Aquarius,
the Fishes, and Canopus, figured by a star of the first
magnitude. To the north is seen near the constellation of
the Swan a large star with seven rays, meant to represent
the brightest of those which compose the Great Bear. The
stars of Cassiopeia are not only misplaced, but roughly
represented. The Lyre is curiously drawn. The positions
of the constellations just named are all wrong in this figure,
just as we find those of towns in maps of the earth. The
cartographers of the middle ages, with incredible ignorance,
misplaced in general every locality. They did the same for
the constellations in the celestial hemispheres. In the
heaven of Jupiter, and in that of Saturn we read the words—
Seraphim, Dominationes, Potestates, Archangeli, Virtutes
coelorum, Principatus, Throni, Cherubim, all derived from
their theology. A veritable muddle! The angels placed
with the heroes of mythology, the immortal virgins with
Venus and Andromeda, and the Saints with the Great Bear,
the Hydra, and the Scorpion !

   Another such richly illuminated manuscript in the library
at Ghent, entitled Liber Floridus, contains a drawing similar
to this under the title *Astrologia secundum Bedum*. Only,
instead of the earth, there is a serpent in the centre with the
name Great Bear, and the twins are represented by a man
and woman, Andromeda in a chasuble, and Venus as a nun !

Several similar ones might be quoted, varying more or less from this; one, executed in a geographical manuscript of the fifteenth century, has the tenth sphere, being that of the fixed stars, then the crystalline heaven, and then the immovable heaven, "which," it says, "according to sacred and certain theology, is the dwelling-place of the blessed, where may we live for ever and ever, Amen;" "this is also called the empyreal heaven." Near each planet the author marks the time of its revolution, but not at all correctly.

The constructors of these systems were not in the least doubt as to their reality, for they actually measured the distance between one sphere and another, though in every case their numbers were far from the truth as we now know it. We may cite as an example an Italian system whose spheres were as follows:—Terra, Aqua, Aria, Fuoco, Luna, Mercurio, Venus, Sol, Marte, Giove, Saturno, Stelle fixe, Sfera nona, Cielo empyreo. Attached to the design is the following table of dimensions which we may copy:—

	Miles.
From the centre of the Earth to the surface . . .	3,245
„ „ „ „ inner side of the heaven of the Moon	107,936
Diameter of Moon	1,896
From the centre of the Earth to Mercury .	209,198
Diameter of Mercury	230
From the centre of the Earth to Venus . . .	579,320
Diameter of Venus	2,884
From the centre of the Earth to the Sun	3,892,866
Diameter of the Sun	35,700

PLATE VII.—HEAVENS OF THE FATHERS.

	Miles.
From the centre of the Earth to Mars	4,268,629
Diameter of Mars	7,572
From the centre of the Earth to Jupiter . .	8,323,520
Diameter of Jupiter	29,641
From the centre of the Earth to outside of Saturn's heaven	52,544,702
Diameter of Saturn	29,202
From the centre of the Earth to the fixed stars	73,387,747

The author states that he cannot pursue his calculations further, and condescends to acknowledge that it is very difficult to know accurately what is the thickness of the ninth and of the crystalline heavens !

Perhaps, however, these reckonings are better than those of the Egyptians, who came to the conclusion that Saturn was only distant 492 miles, the sun only 369, and the moon 246.

These numerous variations and adaptations of the Ptolemaic system, prove what a firm hold it had taken, and how it reigned supreme over all minds. Nor are we merely left to gather this. They consciously looked to Ptolemy as their great light, if we may judge from an emblematic drawing taken from an authoritative astronomical work, the *Margarita Philosophica*, which we give on the opposite page.

In all the systems derived from Ptolemy, the order of the planets remained the same, and Mercury and Venus were placed nearer to the earth than the sun is. According to many authors, however, Plato made a variation in this

respect, by putting them outside the sun, on the ground
that they never were seen to pass across its surface.   He

Fig. 17.

had obviously never heard of the "Transit of Venus."
This arrangement was adopted by Theon, in his commentary

on the *Almagesta* of Ptolemy, and afterwards by Geber,
who alone among the Arabians departed from the strict
Ptolemaic system.

The Egyptians improved upon this idea, and made the first
step towards the true system, by representing these two planets,
Mercury and Venus, as revolving round the sun instead of

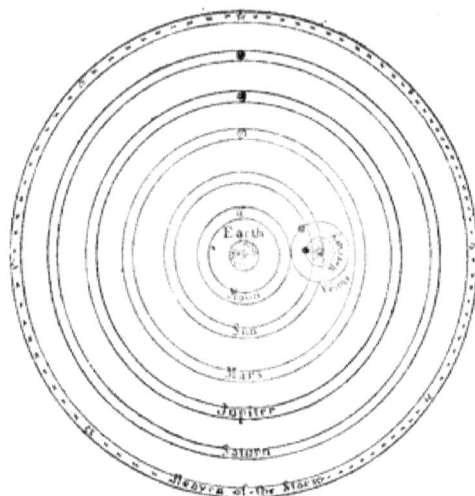

FIG. 18.—EGYPTIAN SYSTEM.

the earth. All the rest of their system was the same as that
of Ptolemy, for the sun itself, and the other planets and the
fixed stars all revolved round the earth in the centre. This
system of course accounted accurately for the motions of the
two inferior planets, whose nearness to the sun may have
suggested their connection with it. This system was in

vogue at the same time as Ptolemy's, and numbers Vitruvius
amongst its supporters.

In the fifth century of our era Martian Capella taught a
variation on the Egyptian system, in which he made Mercury

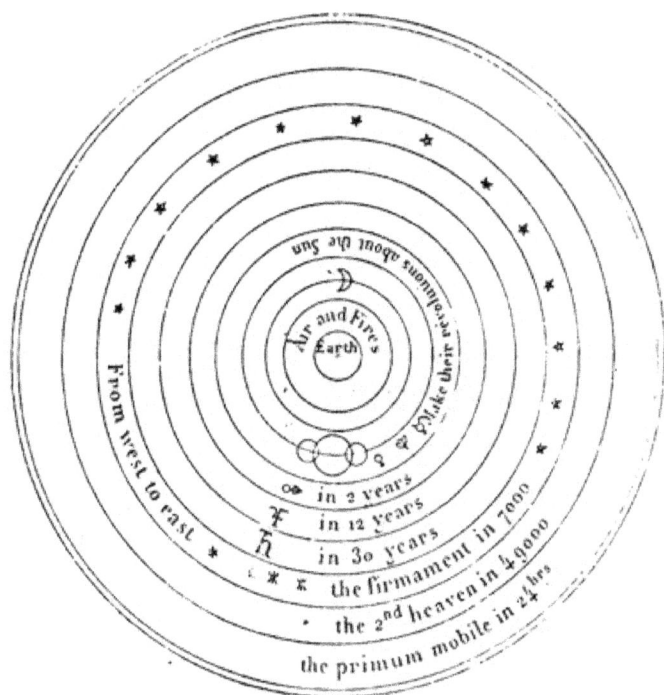

FIG. 19.—CAPELLA'S SYSTEM.

and Venus revolve in the same orbit round the sun. In the
treatise entitled *Quod Tellus non sit Centrum Omnibus
Planetis*, he explains that when Mercury is on this side
of the orbit it is nearer to us than Venus, and farther off

o 2

from us than that planet when it is on the other side. This
hypothesis was also adopted in the middle ages.

We have here indicated the time of the revolution of
the various planets, and notice that the firmament is
said to move round from west to east in 7,000 years ;
the second heaven in 49,000, while the *primum mobile*
outside moved in the contrary direction in twenty-four
hours.

These Egyptian systems survived in some places the
true one, as they were thought to overcome the chief
difficulties of the Ptolemaic without interfering with the
stability of the earth, and they were known as the *common
system*, *i.e.* containing the elements of both.

Such were the astronomical systems in vogue before the
time of Copernicus—all of them based upon the principle of
the earth being the immovable centre of the universe. We
must now turn to trace the history of the introduction of
that system which has completely thrown over all these
former ones, and which every one knows now to be the true
one—the Copernican.

No revolution is accomplished, whether in science or
politics, without having been long in preparation. The
theory of the motion of the earth had been conceived,
discussed, and even taught many ages before the birth of
Copernicus. And the best proof of this is the acknow-
ledgment of Copernicus himself in his great work *De*

Revolutionibus Orbium Cœlestium, in which he laid down the principles of his system. We will quote the passage in which it is contained.

"I have been at the trouble," he writes, "to read over all the works of philosophers that I could procure, to see if I could find in them any different opinion to that which is now taught in the schools respecting the motions of the celestial spheres. And I saw first in Cicero that Mœtas had put forth the opinion that the earth moves. (Mœtam sensisse terram moveri.) Afterwards I found in Plutarch that others had entertained the same idea."

Here Copernicus quotes the original as far as it relates to the system of Philolaus, to the effect "that the earth turns round the region of fire (ethereal region), and runs through the zodiac like the sun and the moon." The principal Pythagoreans, such as Archytas of Tarentum, Heraclides of Pontium, taught also the same doctrine, saying that "the earth is not immovable in the centre of the universe, but revolves in a circle, and is far from occupying the chief place among the celestial bodies."

Pythagoras learnt this doctrine, it is said, from the Egyptians, who in their hieroglyphics represented the symbol of the sun by the stercoral beetle, because this insect forms a ball with the excrement of the oxen, and lying down on its back, turns it round and round with its legs.

Timæus of Locris was more precise than the other
Pythagoreans in calling "the five planets the organs of
time, on account of their revolutions," adding that we must
conclude, that the earth is not immovable in one place,
but that it turns, on the contrary, about itself, and travels
also through space.

Plutarch records that Plato, who had always taught that
the sun turned round the earth, had changed his opinion
towards the end of his life, regretting that he had not
placed the sun in the centre of the universe, which was
the only place, he then thought, that was suitable for
that star.

Three centuries before Jesus Christ, Aristarchus of
Samos is said by Aristotle to have composed a special
work to defend the motion of the earth against the contrary
opinions of philosophers. In this work, which is now lost,
he laid down in the most positive manner that "the sun
remains immovable, and that the Earth moves round it in
a circular curve, of which that star is the centre." It
would be impossible to state this in clearer terms; and
what makes his meaning more clear, if possible, is that he
was persecuted for it, being accused of irreligion and of
troubling the repose of Vesta—"because," says Plutarch,
"in order to explain the phenomena, he taught that the
heavens were immovable, and that the earth accomplished
a motion of translation in an oblique line, at the same

time that it turned round its own axis." This is exactly
the opinion that Copernicus took up, after an interval of
eighteen centuries—and he too was accused of irreligion.

In passing from the Greeks to the Romans, and from
them to the middle ages, the doctrine of Aristarchus
underwent a curious modification, assimilating it to the
system of Tycho Brahe, which we shall hereafter consider,
rather than to that of Copernicus. This consisted in
making the planets move round the sun, while the sun
itself revolved round the earth, and carried them with
him, and the heavens revolved round all. Vitruvius and
Macrobius both taught this doctrine. Although Cicero and
Seneca, with Aristotle and the Stoics, taught the immobility
of the earth in the centre of the universe, the question
seemed undecided, to Seneca at least, who writes:—"It
would be well to examine whether it is the universe that
turns about the immovable earth, or the earth that moves,
while the universe remains at rest. Indeed some men have
taught that the earth is carried along, unknown to ourselves,
that it is not the motion of the heavens that produces the
rising and setting of the stars, but that it is we who rise
and set relatively to them. It is a matter worthy of contemplation, to know in what state we are—whether we are
assigned an immovable or rapidly-moving home—whether
God makes all things revolve round us, or we round them."

The double motion of the earth, then, is an idea revived

from the Grecian philosophers. The theory was known
indeed to Ptolemy, who devotes a whole chapter in his
celebrated *Almagesta* to combat it. From his point of view
it seemed very absurd, and he did not hesitate to call it so;
and it was in reality only when fresh discoveries had altered
the method of examining the question that the absurdities
disappeared, and were transferred to the other side. Not
until it was discovered that the earth was no larger and no
heavier than the other planets could the idea of its revolution and translation have appeared anything else than
absurd. We are apt to laugh at the errors of former great
men, while we forget the scantiness of the knowledge they
then possessed. So it will be instructive to draw attention
to Ptolemy's arguments, that we may see where it is that
new knowledge and ideas have led us, as they would doubtless have led him, had he possessed them, to a different
conclusion.

His argument depends essentially on the observed effects
of weight. "Light bodies," he says, "are carried
towards the circumference, they appear to us to go *up*;
because we so speak of the space that is over our heads,
as far as the surface which appears to surround us. Heavy
bodies tend, on the contrary, towards the middle, as towards
a centre, and they appear to us to fall *down*, because we
so speak of whatever is under our feet, in the direction
of the centre of the earth. These bodies are piled up round

the centre by the opposed forces of their impetus and
friction. We can easily see that the whole mass of the
earth, being so large compared with the bodies that fall upon
it, can receive them without their weight or their velocity
communicating to it any perceptible oscillation. Now if
the earth had a motion in common with all the other heavy
bodies, it would not be long, on account of its weight, in
leaving the animals and other bodies behind it, and without
support, and it would soon itself fall out of heaven. Such
would be the consequences of its motion, which are most
ridiculous even to imagine."

Against the idea of the earth's diurnal rotation he
argued as follows :—"There are some who pretend that
nothing prevents us from supposing that the heaven remains
immovable, and the earth turns round upon its axis from
west to east, accomplishing the rotation each day. It is
true that, as far as the stars are concerned, there is nothing
against our supposing this, if guided only by appearances,
and for greater simplicity ; but those who do so forget how
thoroughly ridiculous it is when we consider what happens
near us and in the air. For even if we admit, which is
not the case, that the lighter bodies have no motion, or
only move as bodies of a contrary nature, although we see
that aërial bodies move with greater velocity than terrestrial
—if we admit that very dense and heavy bodies have
a rapid and constant motion of their own, whereas

in reality they obey but with difficulty the impulses communicated to them—we should then be obliged to assert that the earth, by its rotation, has a more rapid motion than any of the bodies that are round it, as it makes so large a circuit in so short a time. In this case the bodies which are not supported by it would appear to have a motion contrary to it, and no cloud or any flying bird could ever appear to go to the east, since the earth would always move faster than it in that direction."

The *Almagesta* was for a long time the gospel of astronomers; to believe in the motion of the earth was to them more than 'an innovation, it was simply folly. Copernicus himself well expresses the state of opinion in which he found the question, and the process of his own change, in the following words:—"And I too, taking occasion by these testimonies, commenced to cogitate on the motion of the earth, and although that opinion appeared absurd, I thought that as others before me had invented an assemblage of circles to explain the motion of the stars, I might also try if, by supposing the earth to move, I could not find a better account of the motions of the heavenly bodies than that with which we are at present contented. After long researches, I am at last convinced that if we assign to the circulation of the earth the motions of the other planets, calculation and observation will agree better together. And I have no doubt that mathematicians will be of my opinion, if they will take the trouble to consider carefully and not superficially the demonstrations I shall give in this work." Although the opinions of Copernicus had been held before, it is very just that his should be the name by which they are known ; for during the time that elapsed before he wrote, the adherents of such views became fewer and fewer, until at last the very remembrance of them was almost forgotten, and it required research to know who had held them and taught them. It took him thirty years' work to establish them on a firm basis. We shall make no excuse for quoting further from his book, that we may know exactly the circumstances, as far as he tells us, of his giving this system to the world.

" I hesitated for a long time whether I should publish my commentaries on the motions of the heavenly bodies, or whether it would not be better to follow the example of certain Pythagoreans, who left no writings, but communicated the mysteries of their philosophy orally from man to man among their adepts and friends, as is proved by the letter of Lysidas to Hipparchus. They did not do this, as some suppose, from a spirit of jealousy, but in order that weighty questions, studied with great care by illustrious men, might not be disparaged by the idle, who do not care to undertake serious study, unless it be lucrative, or by shallow-minded men, who, though devoting themselves to science, are of so indolent a spirit that they only intrude among philosophers, like drones among bees.

"When I hesitated and held back, my friends pressed me on. The first was Nicolas Schonberg, Cardinal of Capua, a man of great learning. The other was my best friend, Tideman Gysius, Bishop of Culm, who was as well versed in the Holy Scriptures as in the sciences. The latter pressed me so much that he decided me at last to give to the public the work I had kept for more than twenty-seven years. Many illustrious men urged me, in the interest of mathematics, to overcome my repugnance and to let the fruit of my labours see the light. They assured me that the more my theory of the motion of the earth appeared absurd, the more it would be admired when the publication of my work had dissipated doubts by the clearest demonstrations. Yielding to these entreaties, and buoying myself with the same hope, I consented to the printing of my work."

He tried to guard himself against the attacks of dogmatists by saying, " If any evil-advised person should quote against me any texts of Scripture, I deprecate such a rash attempt. Mathematical truths can only be judged by mathematicians."

Notwithstanding this, however, his work, after his death, was condemned by the Index in 1616, under Paul V.

On examining the ancient systems, Copernicus was struck by the want of harmony in the arrangements proposed, and by the arbitrary manner in which new principles were introduced and old ones neglected, comparing the system to a collection of legs and arms not united to any trunk, and it was the

simplicity and harmony which the one idea of the motion of
the earth introduced into the whole system that convinced him
most thoroughly of its truth.

He knew well that new views and truths would appear as
paradoxes, and be rejected by men who were wedded to old
doctrines, and on this account he took such pains to show

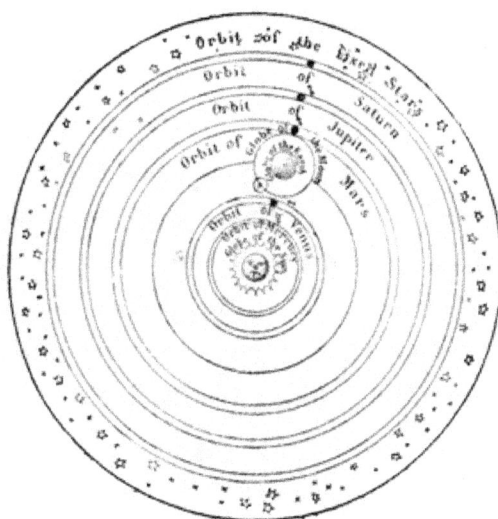

FIG. 20.—THE COPERNICAN SYSTEM.

that these views had been held before, and thus to disarm
them of their apparent novelty.

Copernicus dealt only with the six planets then known and
the sun and moon. As to the stars, he had no idea that they
were suns like our own, at immense and various distances
from us. The knowledge of the magnitude of the sidereal

universe was reserved for our own century, when it was
discovered by the method of parallaxes. We will give
Copernicus's own sketch of the planetary system :—

" In the highest place is the sphere of the fixed stars, an
immovable sphere, which surrounds the whole of the universe.
Among the movable planets the first is Saturn, which requires
thirty years to make its revolution. After it Jupiter accomplishes its journey in twelve years ; Mars follows, requiring two
years. In the fourth line come the earth and the moon which in
the course of one year return to their original position. The
fifth place is occupied by Venus, which requires nine months
for its journey. Mercury occupies the sixth place, whose
orbit is accomplished in eighty days. In the midst of all
is the sun. What man is there, who in this majestic
temple could choose another and better place for that brilliant
lamp which illuminates all the planets with their satellites ?
It is not without reason that the sun is called the lantern
of the world, the soul and thought of the universe. In
placing it in the centre of the planets, as on a regal throne,
we give it the government of the great family of celestial
bodies."

The hypothesis of the motion of the earth in its orbit
appeared simply to Copernicus as a good basis for the exact
determination of the ratios of the distances of the several
planets about the sun. But he did not give up the excentrics
and epicycles for the explanation of the irregular motions of

the planets, and certain imaginary variations in the precession of the equinoxes and the obliquity of the ecliptic.
According to him the earth was endowed with three
different motions, the first about its axis, the second along
the ecliptic, and a third, which he called the declination,
moving it backwards along the signs of the zodiac from east
to west. This last motion was invented to explain the
phenomena of the seasons. He thought, like many other
ancient philosophers, that a body could not turn about
another without being fixed in some way to it—by a crystal
sphere, or something—and in this case that the same surface
would each day be presented to the sun, and so it requires a
third rotation, by which its axis may remain constantly
parallel to itself. Galileo, however, afterwards demonstrated
the independence of the two motions in question, and proved
that the third was unnecessary.

Copernicus was born in the Polish village of Thorn, in 1473,
and died in 1543, at Warmia, of which he was canon, and
where he built an observatory. The voyages of his youth, his
labours, adversities, and old age at last broke him down, and
in the winter of 1542 he took to his bed, and was incapable
of further work. His work, which was just finished printing
at Nuremberg, was brought to him by his friends before he
died. He soon after completely failed in strength, and passed
away tranquilly on the 23rd of May, 1543.

The Copernican system required, however, establishing in

PLATE VIII.—DEATH OF COPERNICUS.

the minds of astronomers generally before it took the
place it now holds, and this work was done by Galileo—a
name as celebrated as that of Copernicus himself, if not
more so. This perhaps is due not only to his demonstra-
tion of the motion of the earth, but to his introduction of
experimental philosophy, and his observational method in
astronomy.

The next advance was made by Kepler, who overthrew at
one blow all the excentrics and epicycles of the ancients,
when by his laborious calculations he proved the ellipticity
of the orbit of Mars.

The Grecian hypotheses were the logical consequences of
two propositions which were universally admitted as axioms
in the early and middle ages. First, that the motions of the
heavenly bodies were uniform ; second, that their orbits
were perfect circles. Nothing appeared more natural than
this belief, though false. So then when Kepler, in 1609,
recognised the fact, by incontestable geometrical measurements, that Mars described an oval orbit round the sun, in
which its velocity varied periodically, he could not believe
either his observation or his calculation, and he puzzled his
brain to discover what secret principle it was that forced the
planet to approach and depart from the sun by turns.
Fortunately for him, in this inquietude he came across a
treatise by Gilbert, *De Magnate,* which had been published in
London nine years before. In this remarkable work Gilbert

P

proved by experiment that the earth acts on magnetized needles and on bars of iron placed near its surface just as a magnet does—and by a conjectural extension of this fact, which was a vague presentiment of the truth, he supposed that the earth itself might be retained in its constant orbit round the sun by a magnetic attraction. This idea was a ray of light to Kepler. It led him to see the secret cause of the alternating motions that had troubled him so much, and in the joy of that discovery he said, " If we find it impossible to attribute the vibration to a magnetic power residing in the sun, acting on the planet without any material medium between, we must conclude that the planet is itself endowed with a kind of intelligent perception which gives it power to know at each instant the proper angles and distances for its motion." In the result Kepler was led to enunciate to the world his three celebrated laws:—

1st. That the planets move in ellipses, of which the sun is in one of the foci.

2nd. The spaces described by the ideal radius which joins each planet to the sun are proportional to the times of their description. In other words, the nearer a planet is to the sun, the faster it moves.

3rd. The squares of the times of revolution are as the cubes of the major axes of the orbits.

Such were the laws of Kepler, the basis of modern astronomy, which led in the hands of Newton to the simple explanation by universal gravitation, which itself is now asking to be explained.

We are not to suppose that the system of Copernicus was universally accepted even by astronomers of note. By some an attempt was made to invent a system which should have all the advantages of this, and yet if possible save the immobility of the earth. Such was that of Tycho Brahe, who was born three years after the death of Copernicus, and died in 1601. He was one of the most laborious and painstaking observers of his time, although by the peculiarity of fate he is known generally only by his false system.

In 1577, Tycho Brahe wrote a little treatise, *Tychonis Brahe, Dani, De Mundi Ætherei Recentioribus phenomenis, à propos* of a comet that had lately appeared. He speaks at length of his system as follows :—" I have remarked that the ancient system of Ptolemy is not at all natural, and too complicated. But neither can I approve of the new one introduced by the great Copernicus after the example of Aristarchus of Samos. This heavy mass of earth, so little fit for motion, could not be displaced in this manner, and moved in three ways, like the celestial bodies, without a shock to the principles of physics. Besides, it is opposed to Scripture ! I think then," he adds, "that we must decidedly and without doubt place the earth immovable in the centre of world, according to the belief of the ancients and the testimony of Scripture.

In my opinion the celestial motions are arranged in such a way that the sun, the moon, and the sphere of the fixed stars, which incloses all, have the earth for their centre. The five planets turn about the sun as about their chief and king, the sun being constantly in the centre of their orbits, and accompany it in its annual motion round the

FIG. 21.—TYCHO BRAHE'S SYSTEM.

earth." This system perfectly accounts for the apparent motions of the planets as seen from the earth, and is essentially a variation on the Copernican, rather than on the Ptolemaic system, but it lent itself less readily to future discoveries. It simply amounts, as far as the solar system is concerned, to impressing upon all the rest of it the motions of the earth, so as to leave the latter at rest; and were the

sun only as large with respect to the earth as it seems, were
the planets really smaller than the moon, and the stars only
at a short distance, and smaller than the planets, it might seem
more natural that they should move than the earth; but when
all these suppositions were disproved, the very argument of
Tycho Brahe for the stability of the earth turned the other way,
and proved as incontestably that it moved. In the Copernican
system, however, these questions are of no consequence; if
the sun be at rest, this mass makes no difference; if the earth
moves like the planets, their relative size does not alter
anything; and if stars are immovable they may be at any
distance and of any magnitude.

The objections of Tycho Brahe to the earth's motion were :
First, that it was too heavy—we know now, however, that
some other planets are heavier—and that the sun, which he
would make move instead, is 340,000 times as heavy.
Secondly, that if the earth moved, all loose things would be
carried from east to west; but we have experience of many
loose things being kept by friction on moving bodies, and
can conceive how, all things may be kept by the attraction of
the earth under the influence of its own motion. Thirdly,
that he could not imagine that the earth was turned upside
down every day, and that for twelve hours our heads are
downwards.

But the existence of the antipodes overcomes this objection,
and shows that there is no up and down in the universe, but

each man calls that *down* which is nearer to the centre of the earth than himself.

A variation on Tycho Brahe's system was attempted by one Longomontanus, who had lived with him for ten years. It consisted in admitting the diurnal rotation, but not the annual revolution, of the earth ; but it made no progress, and was soon forgotten.

More remarkable than this was the attempt by Descartes in the same direction, namely, to hold the principles of Copernicus, and yet to teach the immobility of the earth. His idea of immobility was however very different from that of Tycho Brahe, or of any one else, and would only be called so by those who were bound to believe it at all costs.

His Theory of Vortices, as it is called, will be best given in his own words as contained in his *Les Principes de la Philosophie*, third part, chap. xxvi., entitled, " That the earth is at rest in its heaven, which does not prevent its being carried along with it, and that it is the same with all the planets."

" I adhere," he says, " to the hypothesis of Copernicus, because it seems to me the simplest and clearest. There is no vacuum anywhere in space. . . . The heavens are full of a universal liquid substance. This is an opinion now commonly received among astronomers, because they cannot see how the phenomena can be explained without it. The substance of the heavens has the common property of all liquids, that its

minutest particles are easily moved in any direction, and
when it happens that they all move in one way, they
necessarily carry with them all the bodies they surround,
and which are not prevented from moving by any external
cause. The matter of the heaven in which the planets are
turns round continually like a vortex, which has the Sun for
its centre. The parts that are nearest the Sun move faster
than those that are at a greater distance; and all the planets,
including the earth, remain always suspended in the same
place in the matter of the heaven. And just as in the turns
of rivers, when the water turns back on itself and twists
round in circles, if any twig or light body floats on it, we see
it carry them round, and make them move with it, and even
among these twigs we may see some turning on their own
centre, and those that are nearest to the middle of the
vortex moving quicker than those on the outside; so we may
easily imagine it to be with the planets, and this is all that is
necessary to explain the phenomena. The matter that is round
Saturn takes about thirty years to run its circle; that which
surrounds Jupiter carries it and its satellites round in twelve
years, and so on. . . . The satellites are carried round their
primaries by smaller vortices. . . . The earth is not sustained
by columns, nor suspended in the air by ropes, but it is en-
vironed on all sides by a very liquid heaven. It is at rest, and
has no propulsion or motion, since we do not perceive any in it.
This does not prevent it being carried round by its heaven, and

following its motion without moving itself, just as a vessel

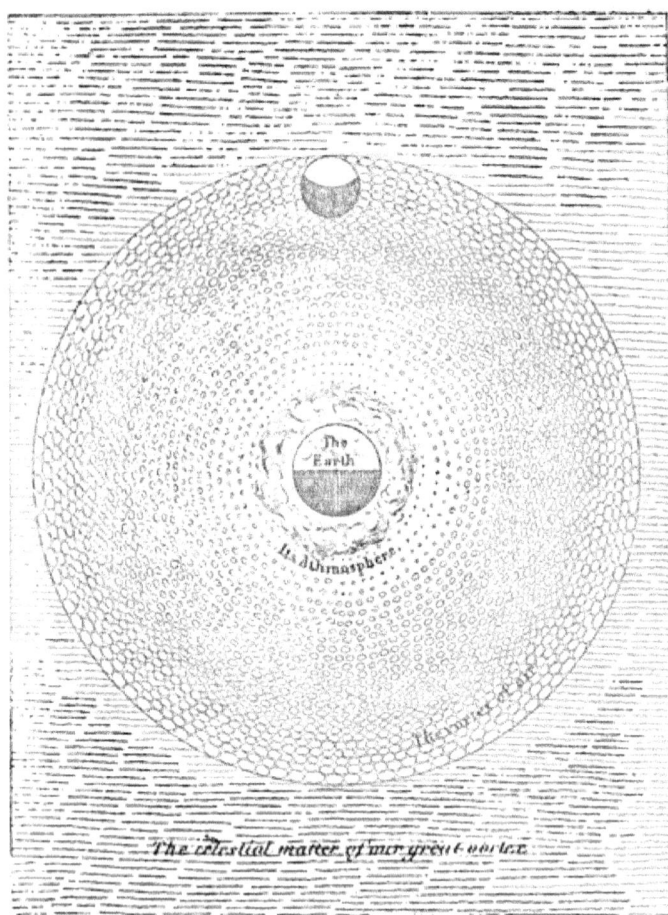

FIG. 22. —DESCARTES' THEORY OF VORTICES.

which is not moved by winds or oars, and is not retained by
anchors, remains in repose in the middle of the sea, although

the flood of the great mass of water carries it insensibly with it. Like the earth, the planets remain at rest in the region of heaven where each one is found. Copernicus made no difficulty in allowing that the earth moves. Tycho, to whom this opinion seemed absurd and unworthy of common sense, wished to correct him, but the earth has far more motion in his hypothesis than in that of Copernicus."

Such is the celebrated theory of vortices. The comparison of the rotation of the earth and planets and their revolution round the sun to the turning of small portions of a rapid stream, may contain an idea yet destined to be developed to account for these motions ; but as used by Descartes it is a mere playing upon words admirably adapted to secure the concurrence of all parties ; those who believed in the motion of the earth seeing that it did not interfere with their ideas in the least, and those who believed in its stability being gratified to find some way by which they might still cling to that belief and yet adopt the new ideas. This was its purpose, and that purpose it well served ; but as a philosophical speculation it was worthless. When former astronomers declared that any planet moved, whether it were the earth or any other, they had no idea of attraction, but supposed the planet fixed to a sphere ; this sphere moving and carrying the planet with it was what they meant by the planet moving : the theory of vortices merely substituted a liquid for a solid sphere, with this disadvantage, that if the

planet were fixed to a solid moving sphere, it *must* move ;
if only placed in a liquid one, that liquid might pass it if it
did not have motion of its own.

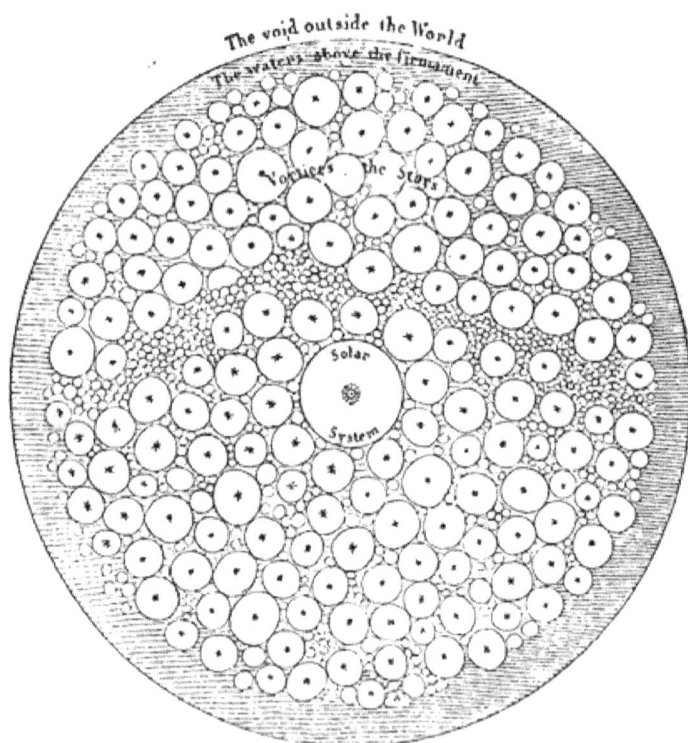

FIG. 23.—VORTICES OF THE STARS.

A variation on Descartes' system of vortices was proposed
in the eighteenth century, which supposed that the sun,
instead of being fixed in the centre of the system, itself

circulated round another centre, carrying Mercury with it.
This motion of the sun was intented to explain the changes
of magnitude of its disc as seen from the earth, and the

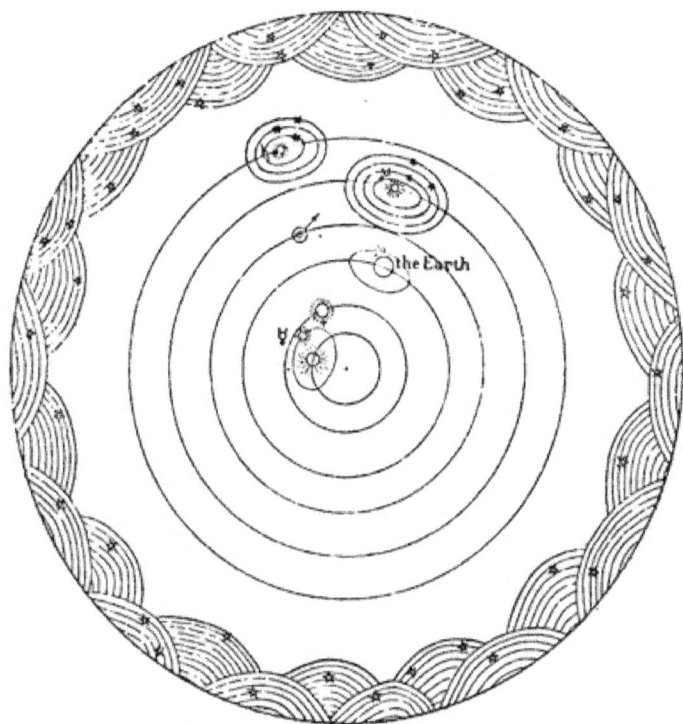

FIG. 24.—VARIATION OF DESCARTES' THEORY.

diurnal and annual variations in its motion, without discard-
ing its circular path.

We have thus noticed all the chief astronomical systems
that have at any time been entertained by astronomers. They

one and all have given way before the universally acknowledged truth about which there is no longer any dispute.
Systems are not now matters of opinion or theory. We
speak of facts as certain as any that can be ascertained in
any branch of knowledge. We have much to learn, but
what we have settled as the basis of our knowledge will
never more be altered as far as we can see.

Of course there have been always fantastic fancies put
forth about the solar system, but they are more amusing
than instructive. Some have said that there is no sun, moon,
or stars, but that they are reflections from an immense light
under the earth. Some savage races say that the moon when
decreasing breaks up into stars, and is renewed each month
by a creative act. The Indians used to say that it was full
of nectar which the gods ate up when it waned, and which
grew again when it waxed. The Brahmins placed the earth
in the centre, and said that the stars moved like fishes in a
sea of liquid. They counted nine planets, of which two are
invisible dragons which cause eclipses; which, since they
happen in various parts of the zodiac, show that these
dragons revolve like the rest. They said the sun was nearer
than the moon, perhaps because it is hotter and brighter.
Berosus the Chaldean gave a very original explanation of the
phases and eclipses of the moon. He said it had one side
bright, and the other side just the colour of the sky, and in
turning it represented the different colours to us.

Before concluding this chapter we may notice what information we possess as to the origin of the names by which the
planets are known. These names have not always been given
to them, and date only from the time when the poets began
to associate the Grecian mythology with astronomy. The
earlier names had reference rather to their several characters,
although there appear to have been among every people two
sets of names applied to them.

The earliest Greek names referred to their various degrees of
brilliancy: thus Saturn, which is not easily distinguished, was
called Phenon, or *that which appears ;* Jupiter was named
Phaëton, *the brilliant ;* Mars was Pysoïs, or *flame-coloured ;*
Mercury, Stilbon, *the sparkling ;* Venus, Phosphorus ; and
Lucifer, *the light-bearer.* They called the latter also Calliste,
the most beautiful. It was also known then as now under the
appellations of the morning star and evening star, indicating
its special position.

With the ancient Accadians, the planets had similar names,
among others. Thus, "Mars was sometimes called *the vanishing
star,* in allusion to its recession from the earth, and Jupiter
the *planet of the ecliptic,* from its neighbourhood to the latter"
(Sayce). The name of Mars raises the interesting question
as to whether they had noticed its phases as well as its
movements—especially when, with reference to Venus, it is
recorded in the "Observations of Bel," that "it rises, and in its
orbit duly grows in size." They had also a rather confusing

system of nomenclature by naming each planet after the star that it happened to be the nearest to at any point of its course round the ecliptic.

Among less cultivated nations also the same practice held, as with the natives of South America, whose name for the sun is a word meaning *it brings the day*; for the moon, *it brings the night*; and for Venus, *it announces the day*.

But even among the Eastern nations, from whom the Greeks and Romans borrowed their astronomical systems, it soon became a practice to associate these planets with the names of the several divinities they worshipped. This was perhaps natural from the adoration they paid to the celestial luminaries themselves on account of their real or supposed influence on terrestrial affairs; and, moreover, as time went on, and heroes had appeared, and they had to find them dwelling-places in the heavens, they would naturally associate them with one or other of the most brilliant and remarkable luminaries, to which they might suppose them translated. Beyond these general remarks, only conjectures can be made why any particular divinity should among the Greeks be connected with the several planets as we now know them. Such conjectures as the following we may make. Thus Jupiter, the largest, would take first rank, and be called after the name of the chief divinity. The soft and sympathising Venus—appearing at the twilight—would well denote the evening star. Mars would receive its name from its red appearance, naturally suggesting carnage and the god of war. Saturn, or Kronos, the god of time, is personified by the slow and almost imperceptible motion of that remote planet. While Mercury, the fiery and quick god of thieves and commerce, is well matched with the hide-and-seek planet which so seldom can be seen, and moves so rapidly.

These were the only planets known to the ancients, and were indeed all that could be discovered without a telescope. If the ancient Babylonians possessed telescopes, as has been conjectured from their speaking, as we have noticed above, of the increase of the size of Venus, and from the finding a crystal lens among the ruins of Nineveh, they did not use them for his purpose.

The other planets now known have a far shorter history. Uranus was discovered by Sir William Herschel on the 13th of March, 1781, and was at first taken for a comet. Herschel proposed to call it Georgium Sidus, after King George III. Lalande suggested it should be named Herschel, after its discoverer, and it bore this name for some time. Afterwards the names, Neptune, Astrœa, Cybele, and Uranus were successively proposed, and the latter, the suggestion of Bode, was ultimately adopted. It is the name of the most ancient of the gods, connected with the then most modern of planets in point of discovery, though also most ancient in formation, if recent theories be correct. Neptune, as everybody knows, was calculated into existence, if one may so speak, by Adams and Leverrier independently, and was first seen, in the quarter indicated, by Dr. Galle at Berlin, in September, 1846, and by universal consent it received the name it now bears.

There are now also known a long series of what are called minor planets, all circulating between Mars and Jupiter, with their irregular orbits inextricably mingled together. Their discovery was led to in a remarkable manner. It was observed that the distances of the several planets might approximately be expressed by the terms of a certain mathematical series, if one term was supplied between Mars and Jupiter—a fact known by the name of Bode's law. When the new planet, Uranus, was found to obey this law, the feeling was so strong that there must be something to represent this missing term, that strong efforts were made to discover it, which led to success, and several, whose names are derived from the minor gods and goddesses, are now well known.

All these planets, like the signs of the zodiac, are indicated by astronomers by certain symbols, which, as they derive their form from the names or nature of the planets, may properly here be explained. The sign of Neptune is ♆, representing the trident of the sea; for Uranus ♅, which is the first letter of Herschel with a little globe below; ♄ is the sickle of time, or Saturn; ♃ is the representation of the first letter of Zeus or Jupiter; ♂ is the lance and buckler of Mars;

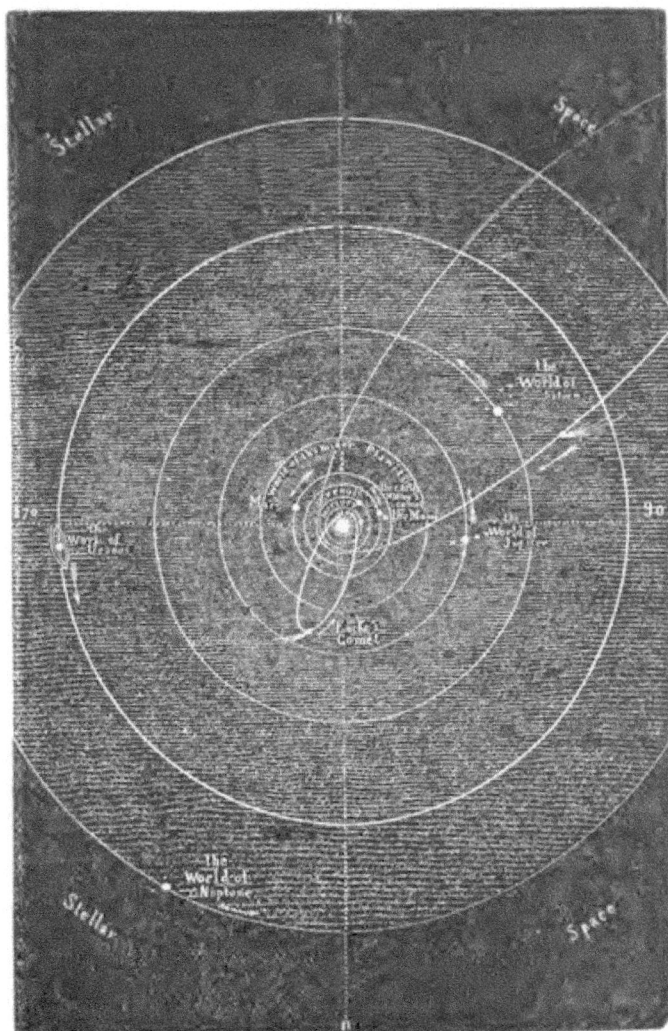

PLATE IX.—THE SOLAR SYSTEM.

♀ the mirror of Venus; ☿ the wand of Mercury; ☉ the sun's disc; and ☽ the crescent of the moon.

The more modern discoveries have, of course, been all made by means of the telescope, and a few words on the history of its discovery may fitly close this chapter.

According to Olbers, a concave and convex lens were first used in combination, to render objects less distant in appearance, in the year 1606. In that year the children of one Jean Lippershey, an optician of Middelburg, in Zealand, were playing with his lenses, and happened to hold one before the other to look at a distant clock. Their great surprise in seeing how near it seemed attracted their father's attention, and he made several experiments with them, at last fixing them as in the modern telescope—in draw tubes. On the 2nd of October, 1606, he made a petition to the States-General of Holland for a patent. The aldermen, however, saw no advantage in it, as you could only look with one eye instead of two. They refused the patent, and though the discovery was soon found of value, Lippershey reaped no benefit.

Galileo was the first to apply the telescope to astronomical observations. He did not have it made in Holland, but constructed it himself on Lippershey's principle. This was in 1609. Its magnifying power was at first 4, and he afterwards increased it to 7, and then to 30. With this he discovered the phases of Venus, the spots on the sun, the four satellites of Jupiter, and the mountains of the moon.

PLATE X.—THE DISCOVERY OF THE TELESCOPE.

Kepler, in 1611, made the first astronomical telescope with two concave glasses.

Huyghens increased the magnifying power successively to 48, 50, and 92, and discovered Saturn's ring and his satellite No. 4.

Cassini, the first director of the Paris Observatory, brought it to 150, aided by Auzout Campani of Rome, and Rives of London. He observed the rotation of Jupiter (1665), that of Venus and Mars (1666), the fifth and third satellites of Saturn (1671), and afterwards the two nearer ones (1684); the other satellites of this planet were discovered, the sixth and seventh, by Sir William Herschel (1789), and the eighth by Bond and Lasel (1848).

We may add here that the satellites of Uranus were discovered, six by Herschel from 1790 to 1794, and two by Lassel in 1851, the latter also discovering Neptune's satellite in 1847.

The rotation of Saturn was discovered by Herschel in 1789, and that of Mercury by Schrœter in 1800.

The earliest telescopes which were reflectors were made by Gregory in 1663 and Newton in 1672. The greatest instruments of our century are that of Herschel, which magnifies 3,000 times, and Lord Rosse's, magnifying 6,000 times, the Foucault telescope at Marseilles, of 4,000, the reflector at Melbourne, of 7,000, and the Newall refractor.

The exact knowledge of the heavens, which makes so grand

PLATE XI.—THE FOUNDATION OF PARIS OBSERVATORY.

a feature in modern science, is due, however, not only to the existence of instruments, but also to the establishment of observatories especially devoted to their use. The first astronomical observatory that was constructed was that at Paris. In 1667 Colbert submitted the designs of it to Louis XIV., and four years afterwards it was completed. The Greenwich Observatory was established in 1676, that of Berlin in 1710, and that of St. Petersburg in 1725. Since then numerous others have been erected, private as well as public, in all parts of the world, and no night passes without numerous observations being taken as part of the ordinary duty of the astronomers attached to them.

CHAPTER IX.

THE TERRESTRIAL WORLD OF THE ANCIENTS.—COSMOGRAPHY AND GEOGRAPHY.

WITH respect to the shape and position of the earth itself in the material universe, and the question of its motion or immobility, we cannot go so far back as in the case of the heavens, since it obviously requires more observation, and is not so pressing for an answer.

Amongst the Greeks several authors appear to have undertaken the subject, but only one complete work has come down to us which undertakes it directly. This is a work attributed to Aristotle, *De Mundo.* It is addressed to Alexander, and by some is considered to be spurious, because it lacks the majestic obscurity that in his acknowledged works repels the reader. Although, however, it is not as obscure as it might be, for the writer, it is quite bad enough, and its dryness and vagueness, its mixture of metaphysical and physical reasoning, logic and observation, and the change that has naturally passed over the meanings of many common

words since they were written, render it very tedious and
unpleasant reading,

Nevertheless, as presenting us with the first recorded ideas
on these questions of the nature and properties of the earth,
it deserves attentive study. It is not a system of observations like those of Ptolemy and the Alexandrian School,
but an entirely theoretical work. It is founded entirely on
logic; but unfortunately, if the premisses are bad, the better
the syllogism the more erroneous will be the conclusion; and
it is just this which we find here. Thus if he be asked
whether the earth turns or the heavens, he will reply that the
earth is *evidently* in repose, and that this is the case not only
because we observe it to be so, but because it is a necessity
that it should be; because repose is *natural* to the earth, and
it is *naturally* in equilibrium. This idea of "natural" leads
very often astray. He is guided to his idea of what is
natural by seeing what is, and then argues that what is, or
appears to be, must be, because it is natural—thus arguing in
a circle. Another example may be given in his answer to
the question, Why must the stars move round the earth? He
says it is natural, because a circle is a more perfect line, and
must therefore be described by the perfect stars, and a circle
is perfect because it has no ends! Unfortunately there are
other curves that have no ends; but the circle was considered,
without more reason, the most perfect curve, and therefore
the planets must move in circles—an idea which had to wait

till Kepler's time to be exploded. One more specimen of
this style may be quoted, namely, his proof that every part
of heaven must be eternally moving, while the earth must be
in the centre and at rest. The proof is this. Everything
which performs any act has been made for the purpose of that
act. Now the work of God is immortality, from which it
follows that all that is divine must have an eternal motion.
But the heavens have a divine quality, and for this reason
they have a spherical shape and move eternally in a circle.
Now when a body has a circular motion, one part of it must
remain at rest in its place, namely, that which is in the
centre; the earth is in the centre—therefore it is at rest.

Aristotle says in this work that there are two kinds of
simple motion, that in a circle and that in a straight line.
The latter belongs to the elements, which either go up or
down, and the former to the celestial bodies, whose nature is
more divine, and which have never been known to change;
and the earth and world must be the only bodies in existence,
for if there were another, it must be the contrary to this, and
there is no contrary to a circle; and again, if there were any
other body, the earth would be attracted towards it, and
move, which it does not. Such is the style of argument
which was in those days thought conclusive, and which with
a little development and inflation of language appeared
intensely profound.

But what brings these speculations to the subject we have

now in hand is this : that when Aristotle thus proves the earth to be immovable in the centre of the universe, he is led on to inquire how it is possible for it to remain in one fixed place. He observed that even a small fragment of earth, when it is raised into the air and then let go, immediately falls without ever stopping in one place—falling, as he supposed, all the quicker according to its weight ; and he was therefore puzzled to know why the whole mass of the earth, notwithstanding its weight, could be kept from falling.

Aristotle examines one by one the answers that have been given to this question. Thus Xenophanes gave to the earth infinitely extended roots, against which Empedocles uses such arguments as we should use now. Thales of Miletus makes the earth rest upon water, without finding anything on which the water itself can rest, or answering the question how it is that the heavier earth can be supported on the lighter water. Anaximenes, Anaxagoras, and Democritus, who make the earth flat, consider it to be sustained by the air, which is accumulated below it, and also presses down upon it like a great coverlet. Aristotle himself says that he agrees with those philosophers who think that the earth is brought to the centre by the primitive rotation of things, and that we may compare it, as Empedocles does, to the water in glasses which are made to turn rapidly, and which does not fall out or move, even though upside down. He also quotes with approval another opinion somewhat similar to this, namely, that of Anaximander, which states that the earth is in repose, on account of its own equilibrium.   Placed in the centre and at an equal distance from its extremities, there is no reason why it should move in one direction rather than the other, and rests immovable in the centre without being able to leave it.

The result of all is that Aristotle concludes that the earth is immovable, in the centre of the universe, and that it is not a star circulating in space like other stars, and that it does not rotate upon its axis ; and he completes the system by stating that the earth is spherical, which is proved by the different aspects of the heavens to a voyager to the north or to the south.

Such was the Aristotelian system, containing far more error than truth, which was the first of any completeness. Scattered ideas, however, on the shape and method of support of the earth and the cause of various phenomena, such as the circulation of the stars, are met with besides in abundance.

The original ideas of the earth were naturally tinged by the prepossessions of each race, every one thinking his own country to be situated in the centre.   Thus among the Hindoos, who lived near the equator, and among the Scandinavians, inhabiting regions nearer the pole, the same meaning attaches to the words by which they express their own country,

medpiama and *medgard,* both meaning the central habitation.
Olympus among the Greeks was made the centre of the
earth, and afterwards the temple. of Delphi.   For the Egyptians the central point was Thebes ; for the Assyrians it was
Babylon ; for the Indians it was the mountain Mero ; for
the Hebrews Jerusalem.   The Chinese always called their
country the central empire.   It was then the custom to
denote the world by a large disc, surrounded on all sides by
a marvellous and inaccessible ocean.   At the extremities of
the earth were placed imaginary regions and fortunate isles,
inhabited by giants or pigmies.   The vault of the sky was
supposed to be supported by enormous mountains and
mysterious columns.

Numerous variations have been suggested on the earliest
supposed form of the earth, which was, as we have seen in a
former chapter, originally supposed to be an immense flat of
infinite depth, and giving support to the heavens.

As travels extended and geography began to be a science.
it was remarked that an immense area of water circumscribed
the solid earth by irregular boundaries—whence the idea of
a universal ocean.

When, however, it was perceived that the horizon at sea
was always circular, it was supposed that the ocean was
bounded, and the whole earth came to be represented as
contained in a circle, beneath which were roots reaching
downwards without end, but with no imagined support.

The Vedic priests asserted that the earth was supported

FIG. 25.—THE EARTH FLOATING.

on twelve columns, which they very ingeniously turned to their own account by asserting that these columns were

FIG. 26.—THE EARTH WITH ROOTS.

supported by virtue of the sacrifices that were made to the

gods, so that if these were not made the earth would collapse.

These pillars were invented in order to account for the passing of the sun beneath the earth after his setting, for which at first they were obliged to imagine a system of

FIG. 27.—THE EARTH OF THE VEDIC PRIESTS.

tunnels, which gradually became enlarged to the intervals between the pillars.

The Hindoos made the hemispherical earth to be supported upon four elephants, and the four elephants to stand on the back of an immense tortoise, which itself floated on the surface of a universal ocean. We are not however to laugh at this as intended to be literal ; the elephants symbolised, it may be, the four elements, or the four directions of the compass, and the tortoise was the symbol for strength and for eternity, which was also sometimes represented by a serpent.

The floating of the earth on water or some other liquid long held ground. It was adopted by Thales, and six centuries

FIG. 28.—HINDOO EARTH.

later Seneca adopts the same opinion, saying that the humid element that supports the earth's disc like a vessel may be either the ocean or some liquid more simple than water.

Diodorus tells us that the Chaldeans considered the earth hollow and boat-shaped—perhaps turned upside down—and this doctrine was introduced into Greece by Heraclitus of Ephesus.

Anaximander represents the earth as a cylinder, the upper face of which alone is inhabited. This cylinder, he states, is one-third as high as its diameter, and it floats freely in the centre of the celestial vault, because there is no reason why it should move to one side rather than the other. Leucippus, Democritus, Heraclitus, and Anaxagoras all adopted this purely imaginary form. Europe made the northern half, and

FIG. 29.—THE EARTH OF ANAXIMANDER.

Lybia (Africa) and Asia the southern, while Delphi was in the centre.

Anaximenes, without giving a precise opinion as to the form of the earth, made it out to be supported on compressed air, though he gave no idea as to how the air was to be compressed.

Plato thought to improve upon these ideas by making the earth cubical. The cube, which is bound by six equal faces, appeared to him the most perfect of solids, and therefore most suitable for the earth, which was to stand in the centre of the universe.

Eudoxus, who in his long voyages throughout Greece and Egypt had seen new constellations appear as he went south, while others to the north disappeared, deduced the sphericity

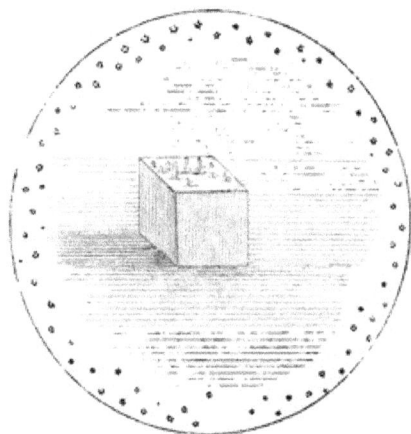

FIG. 30.—PLATO'S CUBICAL EARTH.

of the earth, in which opinion he was followed by Archimedes, and, as we have seen, by Aristotle.

According to Achilles Tatius, Xenophanes gave to the earth the shape of an immense inclined plane, which stretched out to infinity. He drew it in the form of a vast mountain. The summit only was inhabited by men, and round it circulated the stars, and the base was at an infinite depth. Hesiod

R

had before this obscurely said: "The abyss is surrounded by
a brazen barrier; above it rest the roots of the earth."
Epicurus and his school were well pleased with this representation.   If such were the foundations of the earth, then it
was impossible that the sun, and moon, and stars should
complete their revolutions beneath it.   A solid and indefinite
support being once admitted, the Epicurean ideas about the
stars were a necessary consequence; the stars must inevitably
be put out each day in the west, since they are not seen to
return to the place whence they started, and they must be
rekindled some hours afterwards in the east.   In the days of
Augustus, Cleomedes still finds himself obliged to combat
these Epicurean ideas about the setting and rising of the sun
and stars.   "These stupid ideas," he says, "have no other
foundation than an old woman's story—that the Iberians
hear each night the hissing noise made by the burning sun as
it is extinguished, like a hot iron in the waters of the ocean."
Modern travellers have shown us that similar ideas about
the support of the earth have been entertained by more
remote people.   Thus, in the opinion of the Greenlanders,
handed down from antiquity to our own days, the earth is
supported on pillars, which are so consumed by time that
they often crack, and were it not that they are supported by
the incantations of the magicians, they would long since have
broken down.   This idea of the breaking of the pillars may
possibly have originated in the known sinking of the land

beneath the sea, which is still going on even at the present day.

An ancient Egyptian papyrus in the library of Paris gives a very curious hieroglyphical representation of the universe.

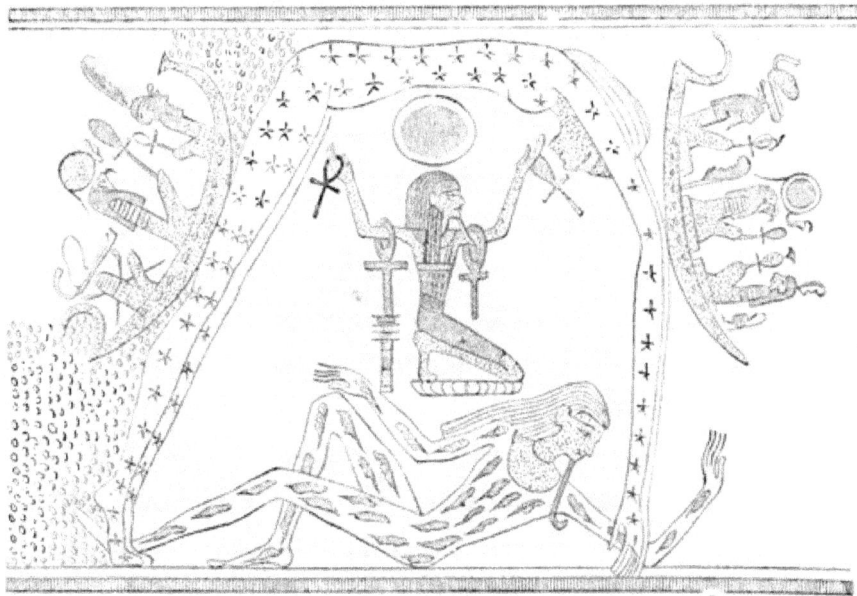

FIG. 31.—EGYPTIAN REPRESENTATION OF THE EARTH.

The earth is here figured under the form of a reclining figure, and is covered with leaves. The heavens are personified by a goddess, which forms the vault by her star-bespangled body, which is elongated in a very peculiar manner. Two boats, carrying, one the rising sun, and the other the setting

R 2

sun, are represented as moving along the heavens over the body of the goddess. In the centre of the picture is the god, Maon, a divine intelligence, which presides over the equilibrium of the universe.

We will now pass on from the early ideas of the general shape and situation of the world to inquire into the first outlines of geographical knowledge of details.

Of all the ancient writings which deal with such questions, the Hebrew Scriptures have the greatest antiquity, and in them are laid down many details of known countries, from which a fair map of the world as known to them might be made out. The prophet Esdras believed that six-sevenths of the earth was dry land—an idea which could not well be exploded till the great oceans had been traversed and America discovered.

More interesting, as being more complete, and written to a certain extent for the very purpose of relating what was known of the geography of the earth, are the writings of the oldest Grecian poets. The first elements of Grecian geography are contained in the two national and almost sacred poems, the *Iliad* and *Odyssey*. So important have these writings been considered in regard to ancient geography, that for many centuries discussions have been carried on with regard to the details, though evidently fictitious, of the voyage of Ulysses, and twenty lines of the *Iliad* have furnished matter for a book of thirty volumes.

The shield of Achilles, forged by Vulcan and described in the eighteenth book of the *Iliad*, gives us an authentic representation of the primitive cosmographical ideas of the age. The earth is there figured as a disc, surrounded on all sides by the *River Ocean*. However strange it may appear to us, to apply the term *river* to the ocean, it occurs too often in Homer and the other ancient poets to admit of a doubt of its being literally understood by them. Hesiod even describes the sources of the ocean at the western extremity of the world, and the representation of these sources was preserved from age to age amongst authors posterior to Homer by nearly a thousand years. Herodotus says plainly that the geographers of his time drew their maps of the world according to the same ideas; the earth was figured with them as a round disc, and the ocean as a river, which washed it on all sides.

The earth's disc, the *orbis terrarum*, was covered according to Homer by a solid vault or firmament, beneath which the stars of the day and night were carried by chariots supported by the clouds. In the morning the sun rose from the eastern ocean, and in the evening it declined into the western; and a vessel of gold, the mysterious work of Vulcan, carried it quickly back by the north, to the east again. Beneath the earth Homer places, not the habitation of the dead, the caverns of Hades, but a vault called Tartarus, corresponding to the firmament. Here lived the Titans, the enemies of the

gods, and no breath of wind, no ray of light, ever penetrated
to this subterranean world. Writers subsequent to Homer
by a century determined even the height of the firmament
and the depth of Tartarus. An anvil, they said, would take
nine days to fall from heaven to earth, and as many more
to fall from earth to the bottom of Tartarus. This estimate
of the height of heaven was of course far too small. If a
body were to fall for nine days and nights, or 777,600 seconds
under the attraction of the earth, it would only pass over
430,500 miles, that is not much more than half as far again
as the moon. A ray of light would only take two seconds to
pass over that distance, whereas it takes eight minutes
to reach us from the sun, and four hours to come from
Neptune—to say nothing of the distance of the stars.

The limits of the world in the Homeric cosmography were
surrounded by obscurity. The columns of which Atlas was
the guardian were supported on unknown foundations, and
disappeared in the systems subsequent to Homer. Beyond
the mysterious boundary where the earth ended and the
heavens began an indefinite chaos spread out—a confused
medley of life and inanity, a gulf where all the elements of
heaven, Tartarus, and earth and sea are mixed together, a
gulf of which the gods themselves are afraid.

Ideas such as these prevailed long after geometers and
astronomers had proved the spherical form of the globe,
and they were revived by the early Christian geographers

and have left their trace even on the common language of
to-day.

The centre of the terrestrial disc was occupied by the
continent and isles of Greece, which in the time of Homer

FIG. 32.—HOMERIC COSMOGRAPHY.

possessed no general name. The centre of Greece passed
therefore for the centre of the whole world; and in Homer's
system it was reckoned to be Olympus in Thessaly, but the
priests of the celebrated Temple of Apollo at Delphi (known

then under the name of Python) gave out a tradition that
that sacred place was the real centre of the habitable world.

The straits which separate Italy from Sicily were so to
speak the vestibule of the fabulous world of Homer. The
threefold ebb and flow, the howling of the monster Scylla,
the whirlpools of Charybdis, the floating rocks—all tell us
that we are quitting here the region of truth. Sicily itself,
although already known under the name of *Trinacria*, was
filled with marvels ; here the flocks of the Sun wandered in
a charming solitude under the guardianship of nymphs ; here
the Cyclops, with one eye only, and the anthropophagous
Lestrigons scared away the traveller from a land that was
otherwise fertile in corn and wine. Two historical races were
placed by Homer in Sicily, namely the *Sicani*, and the *Siceli*,
or *Siculi*.

To the west of Sicily we find ourselves in the midst of a
region of fables. The enchanted islands of Circe and Calypso,
and the floating island of Eolus can no longer be found,
unless we imagine them to have originated, like Graham's
Island in this century, from volcanic eruptions or elevations,
and to have disappeared again by the action of the sea.

The Homeric map of the world terminated towards the
west by two fabulous countries which have given rise to
many traditions among the ancients, and to many discussions
among moderns. Near to the entrance of the ocean, and not
far from the sombre caverns where the dead are congregated,

Ulysses found the *Cimmerians*, "an unhappy people, who, constantly surrounded by thick shadows, never enjoyed the rays of the sun, neither when it mounted the skies, nor when it descended below the earth." Still farther away, and in the ocean itself, and therefore beyond the limits of the earth, beyond the region of winds and seasons, the poet paints for us a Fortunate Land, which he calls *Elysium*, a country where tempests and winter are unknown, where a soft zephyr always blows, and where the elect of Jupiter, snatched from the common lot of mortals, enjoy a perpetual felicity.

Whether these fictions had an allegory for their basis, or were founded on the mistaken notions of voyagers—whether they arose in Greece, or, as the Hebrew etymology of the name Cimmerian might seem to indicate, in the east, or in Phenicia, it is certain that the images they present, transferred to the world of reality, and applied successively to various lands, and confused by contradictory explanations, have singularly embarrassed the progress of geography through many centuries. The Roman travellers thought they recognised the Fortunate Isles in a group to the west of Africa, now known as the Canaries. The philosophical fictions of Plato and Theopompus about Atlantes and Meropis have been long perpetuated in historical theories; though of course it is possible that in the numerous changes that have taken place in the surface of the earth, some ancient vast and populous island may have descended beneath the level of the sea. On the other side, the poetic imagination created the *Hyperboreans*, beyond the regions where the northern winds were generated, and according to a singular kind of meteorology, they believed them for that reason to be protected from the cold winds. Herodotus regrets that he has not been able to discover the least trace of them; he took the trouble to ask for information about them from their neighbours, the *Arimaspes*, a very clear-sighted race, though having but a single eye; but they could not inform him where the Hyperboreans dwelt. The Enchanted Isles, where the Hesperides used to guard the golden fruit, and which the whole of antiquity placed in the west, not far from the Fortunate Isles, are sometimes called Hyperborean by authors well versed in the ancient traditions. It is also in this sense that Sophocles speaks of the Garden of Phœbus, near the vault of heaven, and not far from the *sources of the night, i.e.* of the setting of the sun.

Avienus explains the mild temperature of the Hyperborean country by the temporary proximity of the sun, since, according to the Homeric ideas, it passes during the night by the northern ocean to return to its palace in the east. This ancient tradition was not entirely exploded in the time of Tacitus, who states that on the confines of Germany might be seen the veritable setting of Apollo beyond the water, and he believes that as in the east the sun gives rise to incense and balm by its great proximity to the earth, so in

the regions where it sets it makes the most precious of juices
to transude from the earth and form amber. It is this idea
that is embedded in the fables of amber being the tears of
gold that Apollo shed when he went to the Hyperborean land
to mourn the loss of his son Æsculapius, or by the sisters of
Phaëton, changed into poplars; and it is denoted by the
Greek name for amber, *electron*—a sun-stone. The Grecian
sages, long before the time of Tacitus, said that this very
precious material was an exhalation from the earth that was
produced and hardened by the rays of the sun, which
they thought came nearer to the earth in the west and in
the north.

Florus, in relating the expedition of Decimus Brutus along
the coast of Spain, gives great effect to the Epicurean views
about the sun, by declaring that Brutus only stopped his con-
quests after having witnessed the actual descent of the sun
into the ocean, and having heard with horror the terrible noise
occasioned by its extinction. The ancients also believed that
the sun and the other heavenly bodies were nourished by the
waters—partly the fresh water of the rivers, and partly the
salt water of the sea. Cleanthes gave the reason for the sun
returning towards the equator on reaching the solstices, that
it could not go too far away from the source of its nourishment. Pytheas relates that in the Island of Thule, six days'
journey north of Great Britain, and in all that neighbourhood,
there was no land nor sea nor air, but a compound of all

three, on which the earth and the sea were suspended, and which served to unite together all the parts of the universe, though it was not possible to go into these places, neither on foot nor in ships. Perhaps the ice floating in the frozen seas and the hazy northern atmosphere had been seen by some navigator, and thus gave rise to this idea. As it stands, the history may be perhaps matched by that of the amusing monk who said he had been to the end of the world and had to stoop down, as there was not room to stand between heaven and earth at their junction.

Homer lived in the tenth century before our era. Herodotus, who lived in the fifth, developed the Homeric chart to three times its size. He remarks at the commencement of his book that for several centuries the world has been divided into three parts—Europe, Asia, and Libya; the names given to them being female. The exterior limits of these countries remained in obscurity notwithstanding that those boundaries of them that lay nearest to Greece were clearly defined.

One of the greatest writers on ancient geography was Strabo, whose ideas we will now give an account of. He seems to have been a disciple of Hipparchus in astronomy, though he criticises and contradicts him several times in his geography. He had a just idea of the sphericity of the earth; but considered it as the centre of the universe, and immovable. He takes pains to prove that there is only one inhabited earth—not in this refuting the notion that the moon and stars might have inhabitants, for these he considered to be insignificant meteors nourished by the exhalations of the ocean; but he fought against the fact of there being on this globe any other inhabited part than that known to the ancients.

It is remarkable to notice that the proofs then used by geographers of the sphericity of the earth are just those which we should use now. Thus Strabo says, "The indirect proof is drawn from the centripetal force in general, and the tendency that all bodies have in particular towards a centre of gravity. The direct proof results from the phenomena observed on the sea and in the sky. It is evident, for example, that it is the curvature of the earth that alone prevents the sailor from seeing at a distance the lights that are placed at the ordinary height of the eye, and which must be placed a little higher to become visible even at a greater distance; in the same way, if the eye is a little raised it will see things which previously were hidden." Homer had already made the same remark.

On this globe, representing the world, Strabo and the cosmographers of his time placed the habitable world in a surface which he describes in the following way: "Suppose a great circle, perpendicular to the equator, and passing through the poles to be described about the sphere. It is plain that the surface will be divided by this circle, and by the

equator into four equal parts. The northern and southern
hemispheres contain, each of them, two of these parts.
Now on any one of these quarters of the sphere let us trace
a quadrilateral which shall have for its southern boundary
the half of the equator, for northern boundary a circle marking the commencement of polar cold, and for the other
sides two equal and opposite segments of the circle that
passes through the poles. It is on one such quadrilateral
that the habitable world is placed." He figures it as an
island, because it is surrounded on all sides by the sea.
It is plain that Strabo had a good idea of the nature of
gravity, because he does not distinguish in any way an
upper or a lower hemisphere, and declares that the quadrilateral on which the habitable world is situated may be any
one of the four formed in this way.

The form of the habitable world is that of a " chlamys," or
cloak. This follows, he says, both from geometry and the great
spread of the sea, which, enveloping the land, covers it both
to the east and to the west and reduces it to a shortened
and truncated form of such a figure that its greatest breadth
preserved has only a third of its length. As to the
actual length and breadth, he says, "it measures seventy
thousand stadia in length, and is bounded by a sea whose
immensity and solitude renders it impassable; while the
breadth is less than thirty thousand stadia, and has for
boundaries the double region where the excess of heat on

one side and the excess of cold on the other render it
uninhabitable."

The habitable world was thus much longer from east to
west than it was broad from north to south; from whence
come our terms *longitude*, whose degrees are counted in
the former direction, and *latitude*, reckoned in the latter
direction.

Eratosthenes, and after him Hipparchus, while he gives
larger numbers than the preceding for the dimensions of the
inhabited part of the earth, namely, thirty-eight thousand
stadia of breadth and eighty thousand of length, declares
that physical laws accord with calculations to prove that
the length of the habitable earth must be taken from the
rising to the setting of the sun. This length extends from
the extremity of India to that of Iberia, and the breadth
from the parallel of Ethiopia to that of Ierne.

That the earth is an island, Strabo considers to be
proved by the testimony of our senses. For wherever men
have reached to the extremities of the earth they have
found the sea, and for regions where this has not been
verified it is established by reasoning. Those who have
retraced their steps have not done so because their passage
was barred by any continent, but because their supplies
have run short, and they were afraid of the solitude; the
water always ran freely in front of them.

It is extraordinary that Strabo and the astronomers of

that age, who recognised so clearly the sphericity of the
earth and the real insignificance of mountains, should yet
have supposed the stars to have played so humble a part,
but so it was; and we find Strabo arguing in what we may
call quite the contrary direction. He says, " the larger the
mass of water that is spread round the earth, so much more

FIG. 33.—THE EARTH OF THE LATER GREEKS.

easy is it to conceive how the vapours arising from it are
sufficient to nourish the heavenly bodies."

Among the Latin cosmographers we may here cite one
who flourished in the first century after Christ, Pomponius
Mela, who wrote a treatise, called *De Situ Orbis.* From whatever source, whether traditional or otherwise, he arrived at
the conclusion, he divided the earth into two continents,
our own and that of the Antichthones, which reached to our

IX.] THE TERRESTRIAL WORLD OF THE ANCIENTS. 257

antipodes. This map was in use till the time of Christopher Columbus, who modified it in the matter of the position of this second continent, which till then remained a matter of mystery.

Of those who in ancient times added to the knowledge then possessed of cosmography, we should not omit to mention the name of Pytheas, of Marseilles, who flourished in the fourth century before our era. His chief observations,

Fig. 34.—Pomponius Mela's Cosmography.

however, were not so closely related to geography as to the relation of the earth with the heavenly bodies. By the observation of the gnomon at mid-day on the day of the solstice he determined the obliquity of the ecliptic in his epoch. By the observation of the height of the pole, he discovered that in his time it was not marked by any star, but formed a quadrilateral with three neighbouring stars, β of the little Bear and κ and α of the Dragon.

S

CHAPTER X.

AFTER the writers mentioned in the last chapter a long interval elapsed without any progress being made in the knowledge of the shape or configuration of the earth. From the fall of the Roman Empire, whose colonies themselves gave a certain knowledge of geography, down to the fifteenth century, when the great impetus was given to discovery by the adventurous voyagers of Spain and Portugal, there was nothing but servile copying from ancient authors, who were even misrepresented when they were not understood. Even the peninsula of India was only known by the accounts of Orientals and the writings of the Ancients until the beginning of the fifteenth century. Vague notions, too, were held as to the limits of Africa, and even of Europe and Asia—while of course they knew nothing of America, in spite of their marking on their maps an antichthonal continent to the south.

Denys, the traveller, a Greek writer of the first century, and Priscian, his Latin commentator of the fourth, still maintained the old errors with regard to the earth. According to them the earth is not round, but leaf-shaped; its boundaries are not so arranged as to form everywhere a regular circle. Macrobius, in his system of the world, proves clearly that he had no notion that Africa was continued to the south of Ethiopia, that is of the tenth degree of N. latitude. He thought, like Cleanthus and Crates and other ancient authors, that the regions that lay nearest the tropics, and were burnt by the sun, could not be inhabited ; and that the equatorial regions were occupied by the ocean. He divided the hemisphere into five zones, of which only two were habitable. "One of them," he said, "is occupied by us, and the other by men of whose nature we are ignorant."

Orosus, writing in the same century (fourth), and whose work exercised so great an influence on the cosmographers of the middle ages and on those who made the maps of the world during that long period, was ignorant of the form or boundaries of Africa, and of the contours of the peninsulas of Southern Asia. He made the heavens rest upon the earth.

S. Basil, also of the fourth century, placed the firmament on the earth, and on this heaven a second, whose upper surface was flat, notwithstanding that the inner surface

s 2

which is turned towards us is in the form of a vault; and he explains in this way how the waters can be held there. S. Cyril shows how useful this reservoir of water is to the life of men and of plants.

Diodorus, Bishop of Tarsus, in the same century, also divided the world into two stages, and compared it to a tent. Severianus, Bishop of Gabala, about the same time, compared the world to a house of which the earth is the ground floor, the lower heavens the ceiling, and the upper, or heaven of heavens, the roof. This double heaven was also admitted by Eusebius of Cæsaræa.

In the fifth, sixth, and seventh centuries science made no progress whatever. It was still taught that there were limits to the ocean. Thus Lactantius asserted that there could not be inhabitants beyond the line of the tropics. This Father of the Church considered it a monstrous opinion that the earth is round, that the heavens turn about it, and that all parts of the earth are inhabited. "There are some people," he says, "so extravagant as to persuade themselves that there are men who have their heads downwards and their feet upwards; that all that lies down here is hung up there; that the trees and herbs grow downwards; and that the snow and hail fall upwards. . . . Those people who maintain such opinions do so for no other purpose than to amuse themselves by disputation, and to show their spirit; otherwise it would be easy to prove by invincible

argument that it is impossible for the heavens to be underneath the earth." (Divine Institution). Saint Augustin also,
in his *City of God*, says : " There is no reason to believe
in that fabulous hypothesis of the antipodes, that is to
say, of men who inhabit the other side of the earth—where
the sun rises when it sets with us, and who have their
feet opposed to ours." . . . . " But even if it were demonstrated by any argument that the earth and world have a
spherical form, it would be too absurd to pretend that any
hardy voyagers, after having traversed the immensity of
the ocean, had been able to reach that part of the world
and there implant a detached branch of the primæval
human family."

In the same strain wrote S. Basil, S. Ambrose, S. Justin
Martyr, S. Chrysostom, Procopius of Gaza, Severianus,
Diodorus Bishop of Tarsus, and the greater number of the
thinkers of that epoch.

Eusebius of Cæsarea was bold enough on one occasion
to write in his Commentaries on the Psalms, that,
" according to the opinion of some the earth is round ; "
but he draws back in another work from so rash an asser-
tion. Even in the fifteenth century the monks of Salamanca
and Alcala opposed the old arguments against the antipodes
to all the theories of Columbus.

In the middle of the sixteenth century Gregory of Tours
adopted also the opinion that the intertropical zone was

uninhabitable, and, like other historians, he taught that the
Nile came from the unknown land in the east, descended

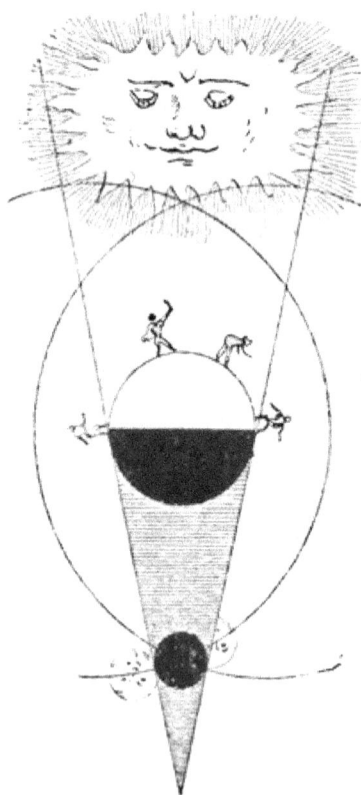

FIG. 35. —THE EARTH'S SHADOW.

to the south, crossed the ocean which separated the antich-
thone from Africa, and then alone became visible.   The
geographical and cosmographical ideas that were then preva- ᴠ

lent may also be judged of by what S. Avitus, a Latin poet
of the sixth century and nephew of the Emperor Flavius
Avitus, says in his poem on the Creation, where he describes
the terrestrial Paradise. " Beyond India," he writes, " *where
the world commences*, where the confines of heaven and earth
are joined, is an exalted asylum, inaccessible to mortals,
and closed by eternal barriers, since the first sin was
committed."

In a treatise on astronomy, published a little after this
in 1581, by Apian and Gemma Frison, they very distinctly
state their belief in a round earth, though they do not go into
details of its surface. The argument is the old one from
eclipses, but the figures they give in illustration are very
amusing, with three or four men of the size of the moon
disporting themselves on the earth's surface. As, however,
they all have their feet to the globe representing the earth,
and consequently have their feet in opposite directions at
the antipodes, the idea is very clearly shown.

" If," they say, " the earth were square, its shadow on the
moon would be square also.

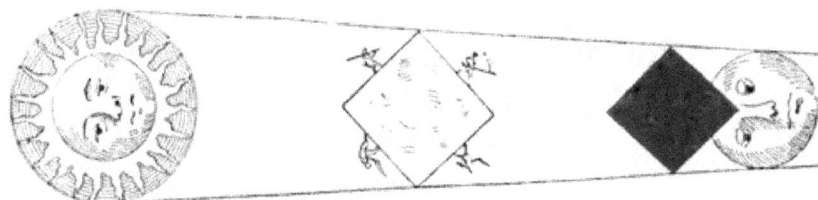

Fig. 36.

"If the earth were triangular, its shadow, during an eclipse of the moon, would also be triangular.

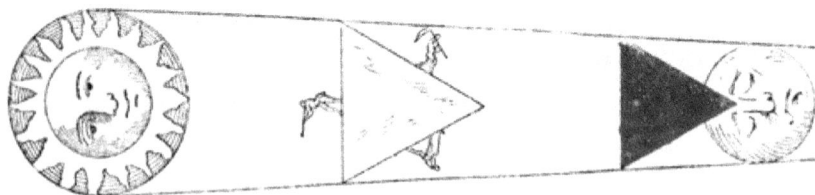

FIG. 37.

"If the earth had six sides, its shadow would have the same figure.

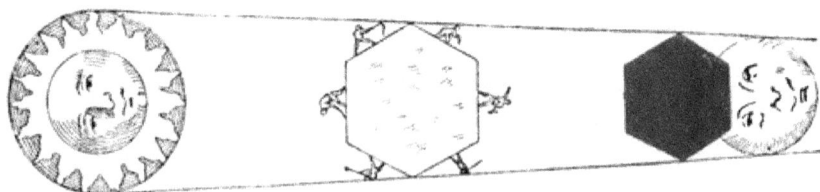

FIG. 38.

"Since, then, the shadow of the earth is round, it is a proof that the earth is round also."

This of course is one of the proofs that would be employed in the present day for the same purpose.

The most remarkable of all the fantastical systems, however, the *chef d'œuvre* of the cosmography of that age, was the famous system of the square earth, with solid walls for supporting the heavens.   Its author was *Cosmas*, surnamed *Indicopleustes* after his voyage to India and Ethiopia.   He was at first a merchant, and afterwards a monk.   He died in 550.   His manuscript was entitled " Christian Topography," and was written in 535.   It was with the object of refuting the opinions of those who gave a spherical form to the earth that Cosmas composed his work after the systems of the Church Fathers, and in opposition to the cosmography of the Gentiles.   He reduced to a systematic form the opinions of the Fathers, and undertook to explain all the phenomena of the heavens in accordance with the Scriptures.   In his first book he refutes the opinion of the sphericity of the earth, which he regarded as a heresy.   In the second he expounds his own system, and the fifth to the ninth he devotes to the courses of the stars.   This mongrel composition is a singular mixture of the doctrines of the Indians, Chaldeans, Greeks, and Christian Fathers.

With respect to his opponents he says, " There are on all sides vigorous attacks against the Church," and accuses them of misunderstanding Scripture, being misled by the eclipses of the sun and moon.   He makes great fun of the idea of rain falling upwards, and yet accuses his opponents of making the earth at the same time the centre and the base of the universe.   The zeal with which these

pretended refutations are used proves, no doubt, that in
the sixth century there were some men, more sensible and
better instructed than others, who preserved the deposit of
progress accomplished by the Grecian genius in the
Alexandrian school, and defended the labours of Hipparchus and Ptolemy ; while it is manifest that the greater
number of their contemporaries kept the old Indian and
Homeric traditions, which were easier to understand, and
more accessible to the false witness of the senses, and not
improved by combination with texts of Scripture misinter-
preted. In fact, cosmographical science in the general
opinion retrograded instead of advancing.

According to Cosmas and his map of the world, the
habitable earth is a plane surface. But instead of being
supposed, as in the time of Thales, to be a disc, he represented it in the form of a parallelogram, whose long sides
are twice the shorter ones, so that man is on the earth
like a bird in a cage. This parallelogram is surrounded
by the ocean, which breaks in in four great gulfs, namely,
the Mediterranean and Caspian seas, and the Persian and
Arabian gulfs.

Beyond the ocean in every direction there exists another
continent which cannot be reached by man, but of which
one part was once inhabited by him before the Deluge. To
the east, just as in other maps of the world, and in later
systems, he placed the *Terrestrial Paradise*, and the four

rivers that watered Eden, which come by subterranean channels to water the post-diluvian earth.

After the Fall, Adam was driven from Paradise; but he and his descendants remained on its coasts until the Deluge carried the ark of Noah to our present earth.

On the four outsides of the earth rise four perpendicular walls, which surround it, and join together at the top in a vault, the heavens forming the cupola of this singular edifice.

The world, according to Cosmas, was therefore a large oblong box, and it was divided into two parts; the first, the abode of men, reaches from the earth to the firmament, above which the stars accomplish their revolutions; there dwell the angels, who cannot go any higher. The second reaches upwards from the firmament to the upper vault, which crowns and terminates the world. On this firmament rest the waters of the heavens.

Cosmas justifies this system by declaring that, according to the doctrine of the Fathers and the Commentators on the Bible, the earth has the form of the Tabernacle that Moses erected in the desert; which was like an oblong box, twice as long as broad. But we may find other similarities,—for this land beyond the ocean recalls the Atlantic of the ancients, and the Mahomedans, and Orientals in general, say that the earth is surrounded by a 'high mountain, which is a similar idea to the walls of Cosmas.

"God," he says, "in creating the earth, rested it on nothing. The earth is therefore sustained by the power of God, the Creator of all things, supporting all things by the word of His power. If below the earth, or outside of it,

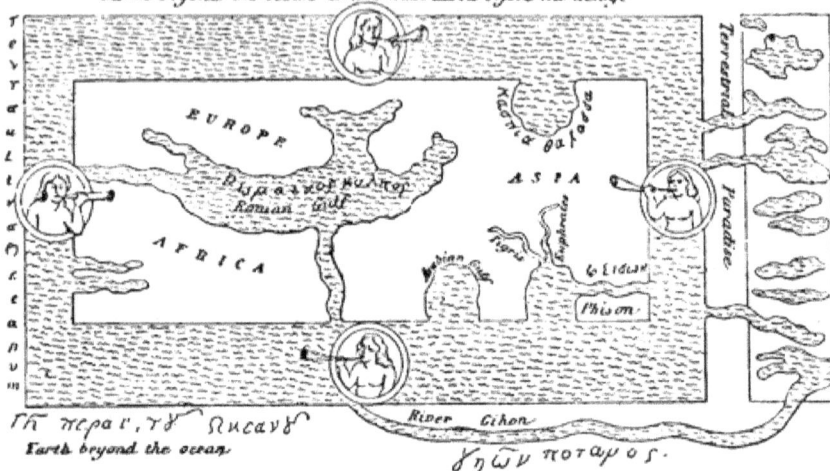

FIG. 39.—THE COSMOGRAPHY OF COSMAS.

anything existed, it would fall of its own accord. So God made the earth the base of the universe, and ordained that it should sustain itself by its own proper gravity."

After having made a great square box of the universe, it remained for him to explain the celestial phenomena, such as the succession of days and nights and the vicissitudes of the seasons.

This is the remarkable explanation he gives. He says that the earth, that is, the oblong table circumscribed on all sides by high walls, is divided into three parts; first the habitable earth, which occupies the middle; secondly, the ocean which surrounds this on all sides; and thirdly, another

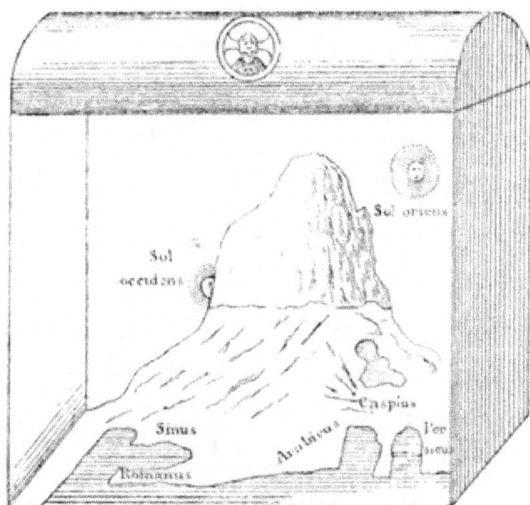

FIG. 40.—THE SQUARE EARTH.

dry land which surrounds the ocean, terminated itself by these high walls on which the firmament rests. According to him the habitable earth is always higher as we go north, so that southern countries are always much lower than northern. For this reason, he says, the Tigris and Euphrates, which run towards the south, are much more rapid than the Nile, which runs northwards. At the extreme north there is a large conical mountain, behind which the sun, moon, planets, and comets all set. These stars never pass below the earth, they only pass behind this great mountain, which hides them for a longer or shorter time from our observation. According as the sun departs from or approaches the north, and consequently is lower or higher in the heavens, he disappears at a point nearer to or further from the base of the mountain, and so is behind it a longer or shorter time, whence the inequality of the days and nights, the vicissitudes of the seasons, eclipses, and other phenomena. This idea is not peculiar to Cosmas, for according to the Indians, the mountain of Someirat is in the centre of the earth, and when the sun appears to set, he is really only hiding behind this mountain.

His idea, too, of the manner in which the motions are performed is strange, but may be matched elsewhere. "All the stars are created," he says, "to regulate the days and nights, the months and the years, and they move, not at all by the motion of the heaven itself, but by the action of certain divine Beings, or *lampadophores.* God made the angels for His service, and He has charged some of them with the motion of the air, others with that of the sun, or the moon, or the other stars, and others again with the collecting of clouds and preparing the rain."

Similar to this were the ideas of other doctors of the

Church, such as S. Hilary and Theodorus, some of whom supposed that the angels carried the stars on their shoulders like the *omophores* of the Manichees; others that they rolled them in front of them or drew them behind; while the

FIG. 41.—EXPLANATION OF SUNRISE.

Jesuit Riccioli, who made astronomical observations, remarks that each angel that pushes a star takes great care to observe what the others are doing, so that the relative distances between the stars may always remain what they ought to be. The Abbot Trithemus gives the exact

succession of the seven angels or spirits of the planets, who
take it in turns during a cycle of three hundred and fifty-
four years to govern the celestial motions from the creation
to the year 1522. The system thus introduced seems to
have been spread abroad, and to have lingered even into the
nineteenth century among the Arabs. A guide of that
nationality hired at Cairo in 1830, remarked to two travellers
how the earth had been made square and covered with
stones, but the stones had been thrown into the four corners,
now called France, Italy, England, and Russia, while the
centre, forming a circle round Mount Sinai, had been given
to the Arabians.

Alongside of this system of the square was another equally
curious—that of the egg. Its author was the famous Venerable Bede, one of the most enlightened men of his time, who
was educated at the University of Armagh, which produced
Alfred and Alcuin. He says: "The earth is an element
placed in the middle of the world, as the yolk is in the
middle of an egg; around it is the water, like the white
surrounding the yolk; outside that is the air, like the mem-
brane of the egg; and round all is the fire which closes it in
as the shell does. The earth being thus in the centre receives
every weight upon itself, and though by its nature it is cold
and dry in its different parts, it acquires accidentally
different qualities; for the portion which is exposed to the
torrid action of the air is burnt by the sun, and is

uninhabitable; its two extremities are too cold to be inhabited, but the portion that lies in the temperate region
of the atmosphere is habitable.   The ocean, which surrounds
it by its waves as far as the horizon, divides it into two parts,
the upper of which is inhabited by us, while the lower is

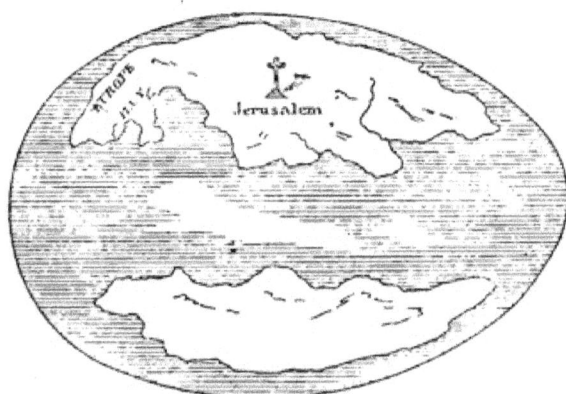

FIG. 42.—THE EARTH AS AN EGG.

inhabited by our antipodes; although not one of them can
come to us, nor one of us to them."

This last sentence shows that however far he may have
been from the truth, he did not, like so many of his contemporaries, stumble over the idea of up and down in the
universe, and so consider the notion of antipodes absurd.

A great number of the maps of the world of the period
followed this idea, and drew the world in the shape of an egg
at rest.   It was broached, however, in another form by

T

Edrisi, an Arabian geographer of the eleventh century, who,
with many others, considered the earth to be like an egg
with one half plunged into the water. The regularity of the
surface is only interrupted by valleys and mountains. He

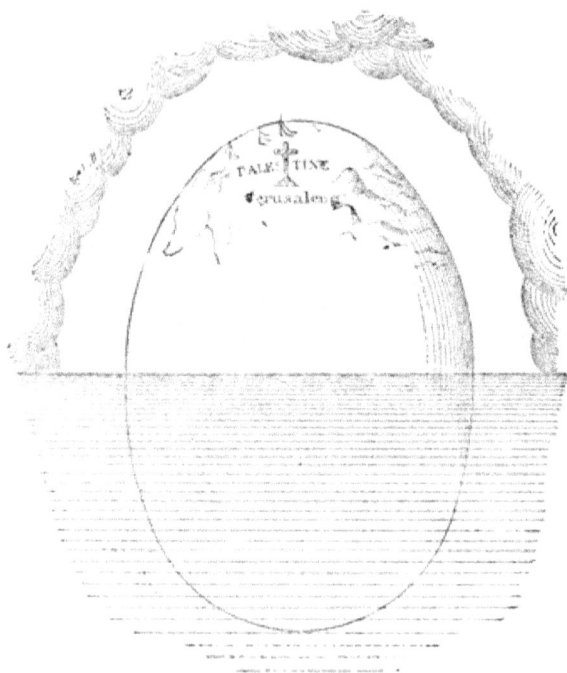

FIG. 43.—THE EARTH AS A FLOATING EGG.

adopted the system of the ancients, who supposed that the
torrid zone was uninhabited. According to him the known
world only forms a single half of the egg, the greater part of
the water belonging to the surrounding ocean, in the midst of

which earth floats like an egg in a basin. Several artists and map-makers adopted this theory in the geographical representations, and so, whether in this way or the last, the egg has had the privilege of representing the form of the earth for nearly a thousand years.

The celebrated Raban Maur, of Mayence, composed in the ninth century a treatise, entitled *De Universo*, divided into twenty-two books. It is a kind of encyclopædia, in which he gives an abridged view of all the sciences. According to his cosmographic system the earth is in the form of a wheel, and is placed in the middle of the universe, being surrounded by the ocean; on the north it is bounded by the Caucasus, which he supposes to be mountains of gold, which no one can reach because of dragons, and griffins, and men of monstrous shape that dwell there. He also places Jerusalem in the centre of the earth.

The treatise of Honorus, entitled *Imago Mundi*, and many other authors of the same kind, represent, 1st, the terrestrial paradise in the most easterly portion of the world, in a locality inaccessible to man; 2nd, the four rivers which had their sources in Paradise; 3rd, the torrid zone, uninhabited; 4th, fantastic islands, transformed from the Atlantis into *Antillia*.

In a manuscript commentary on the Apocalypse, which is in the library of Turin, is a very curious chart, referred to the tenth, but belonging possibly to the eighth century. It

represents the earth as a circular planisphere. The four
sides of the earth are each accompanied by a figure of a
wind, as a horse on a bellows, from which air is poured out,

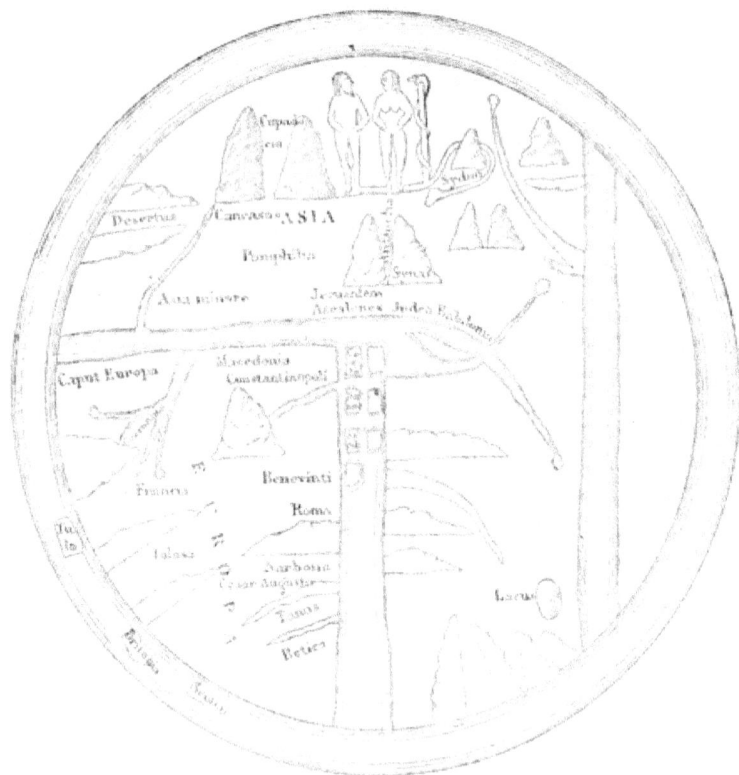

FIG. 44.—EIGHTH CENTURY MAP OF THE WORLD.

as well as from a shell in his mouth. Above, or to the east,
are Adam and Eve with the serpent. To their right is
Asia with two very elevated mountains—Cappadocia and

Caucasus. From thence comes the river *Eusis*, and the sea into which it falls forms an arm of the ocean which surrounds the earth. This arm joins the Mediterranean, and separates Europe from Asia. Towards the middle is Jerusalem, with two curious arms of the sea running past it; while to the south there is a long and straight sea in an east and west direction. The various islands of the Mediterranean are put in a square patch, and Rome, France, and Germany are indicated, while Thula, Britannia, and Scotia are marked as islands in the north-west of the ocean that surrounds the whole world.

We figure below two very curious maps of the world of the tenth century—one of which is round, the other square.

FIG. 45.—TENTH CENTURY MAPS.

The first is divided into three triangles; that of the east, or Asia, is marked with the name of *Shem* ; that of the north,

or Europe, with that of *Japhet;* that of the south, or
Africa, with that of *Cham.* The second is also divided
between the three sons of Noah; the ocean surrounds it,
the Mediterranean forms the upright portion of a cross of
water which divides the Adamic world.

Omons, the author of a geographical poem entitled *The
Image of the World,* composed in 1265, who was called the
Lucretius of the thirteenth century, was not more advanced
than the cosmographers of the former centuries of which
we have hitherto spoken. The cosmographical part of his
poem is borrowed from the system of Pythagoras and the
Venerable Bede. He maintains that the earth is enveloped
in the heavens, as the yoke in the white of an egg, and that
it is in the middle as the centre is within the circle, and he
speaks like Pythagoras of the harmony of the celestial
spheres.

Omons supposed also that in his time the terrestrial
paradise was still existing in the east, with its tree of life,
its four rivers, and its angel with a flaming sword. He
appears to have confounded Hecla with the purgatory of St.
Patrick, and he places the latter in Iceland, saying that it
never ceases to burn. The volcanoes were only, according to
him, the breathing places or mouths of the infernal regions.
The latter he placed with other cosmographers in the
centre of the earth.

Another author, Nicephorus Blemmyde, a monk who

lived during the same century, composed three cosmographical works, among them the following : *On the Heavens
and the Earth, On the Sun and Moon, the Stars, and
Times and Days.* According to his system the earth is
flat, and he adopts the Homeric theory of the ocean
surrounding the world, and that of the seven climates.

Nicolas of Oresmus, a celebrated cosmographer of the
fourteenth century, although his celebrity as a mathematician
attracted the attention of King John of France, who made
him tutor to his son Charles V., was not wiser than those we
have enumerated above. He composed among other works
a *Treatise on the Sphere.* He rejected the theory of an
antichthonal continent as contrary to the faith. A map of
the world, prepared by him about the year 1377, represents
the earth as round, with one hemisphere only inhabited,
the other, or lower one, being plunged in the water. He
seems to have been led by various borrowed ideas, as, for
instance, theological ones, such as the statement in the
Psalms that God had founded the earth upon the waters,
and Grecian ones borrowed from the school of Thales, and
the theories of the Arabian geographers. In fact we have
seen that Edrisi thought that half of the earth was in the
water, and Aboulfeda thought the same. The earth was
placed by Nicolas in the centre of the universe, which he
represented by painting the sky blue, and dotting it over
with stars in gold.

Leonardo Dati, who composed a geographical poem entitled *Della Spera*, during this century, advanced no further. A coloured planisphere showed the earth in the centre of the universe surrounded by the ocean, then the air, then the circles of the planets after the Ptolemaic system, and in another representation of the same kind he figures the infernal regions in the centre of the earth, and gives its diameter as seven thousand miles. He proves himself not to have known one half of the globe by his statement of the shape of the earth—that it is like a **T** inside an **O**. This is a comparison given in many maps of the world in the middle ages, the mean parallel being about the 36th degree of north latitude, that is to say at the Straits of Gibraltar; the Mediterranean is thus placed so as to divide the earth into two equal parts.

John Beauvau, Bishop of Angiers under Louis XI., expresses his ideas as follows:—

"The earth is situated and rests in the middle of the firmament, as the centre or point is in the middle of a circle. Of the whole earth mentioned above only one quarter is inhabited. The earth is divided into four parts, as an apple is divided through the centre by cutting it lengthways and across. If one part of such an apple is taken and peeled, and the peel is spread out over anything flat, such as the palm of the hand, then it resembles the

habitable earth, one side of which is called the east, and
the other the west."

The Arabians adopted not only the ideas of the ancients,
but also the fundamental notions of the cosmographical
system of the Greeks. Some of them, as *Bakony*, regarded
the earth as a flat surface, like a table, others as a ball, of
which one half is cut off, others as a complete revolving
ball, and others that it was hollow within. Others again
went as far as to say that there were several suns and
moons for the several parts of the earth.

In a map, preserved in the library at Cambridge, by Henry,
Canon of St. Marie of Mayence, the form of the world
is given after Herodotus. The four cardinal points are
indicated, and the orientation is that of nearly all the carto-
graphic monuments of the middle ages, namely, the east at
the top of the map. The four cardinal points are four angels,
one foot placed on the disc of the earth; the colours of
their vestments are symbolical. The angel placed at the
Boreal extremity of the earth, or to the north of the
Scythians, points with his finger to people enclosed in the
ramparts of Gog and Magog, *gens immunda* as the legend
says. In his left hand he holds a die to indicate, no doubt,
that there are shut up the Jews who cast lots for the clothes
of Christ. His vestments are green, his mantle and his wings
are red. The angel placed to the left of Paradise has a
green mantle and wings, and red vestments. In his left

hand he holds a kind of palm, and by the right he seems
to mark the way to Paradise. The position of the other
angels placed at the west of the world is different. They
seem occupied in stopping the passage beyond the *Columns*
(that is, the entrance to the Atlantic Ocean). All of them
have golden aureolas. The surrounding ocean is painted of
a clear green.

Another remarkable map of the world is that of Andrea
Bianco. In it we see Eden at the top, which represents the
east, and the four rivers are running out of it. Much of
Europe is indicated, including Spain, Paris, Sweden, Norway,
Ireland, which are named, England, Iceland, Spitzbergen, &c.,
which are not named. The portion round the North Pole
to the left is indicated as "cold beneath the Pole star." In
these maps the systematic theories of the ancient geographers
seem mixed with the doctrines of the Fathers of the Church.
They place generally in the Red Sea some mark denoting the
passage of the Hebrews, the terrestrial paradise at the
extreme east, and Jerusalem in the centre. The towns are
figured often by edifices, as in the list of Theodosius, but
without any regard to their respective positions. Each town
is ordinarily represented by two towers, but the principal
ones are distinguished by a little wall that appears between
these two towers, on which are painted several windows, or
else they may be known by the size of the edifices. St.
James of Compostella in Gallicia and Rome are represented

by edifices of considerable size, as are Nazareth, Troy,
Antioch, Damascus, Babylon, and Nineveh.

One of the most remarkable monuments of the geography
of the last centuries of the middle ages is the map in

FIG. 46.—THE MAP OF ANDREA BIANCO.

Hereford Cathedral, by Richard of Haldingham, not only on
account of its numerous legends, but because of its large
dimensions, being several square yards in area.

On the upper part of this map is represented the Last
Judgment; Jesus Christ, with raised arms, holding in His
hands a scroll with these words, *Ecce testimonium meum.*
At His side two angels carry in their hands the instruments
of His passion.   On the right hand stands an angel with a
trumpet to his mouth, out of which come these words,
Levez si vendres vous par.   An angel brings forward a bishop
by the hand, behind whom is a king, followed by other
personages; the angel introduces them by a door formed of
two columns, which seems to serve as an entrance to an
edifice.

The Virgin is kneeling at the feet of her Son.   Behind her
is another woman kneeling, who holds a crown, which she
seems ready to place on the head of the Mother of Christ,
and by the side of the woman is a kneeling angel, who
appears to be supporting the maternal intercessor.   The
Virgin uncovers her breast and pronounces the words of a
scroll which is held by an angel kneeling in front of her,
*Vei i l' fiz mon piz de deviz lauele chare preistes—Eles mame
lettes dont leit de Virgin qui estes—Syes merci de tous si com
nos mesmes deistes.--R . . . em . . . ont servi kant sauveresse me
feistes.*

To the left another angel, also with a trumpet to his mouth,
gives out the following words, which are written on a scroll,
Leves si alles all fu de enfer estable. A gate, drawn like that of
the entrance, represents probably the passage by which those

must go out who are condemned to eternal pains. In fact the
devil is seen dragging after him a crowd of men, who are tied
by a cord which he holds in his hand.

The map itself commences at its upper part, that is, the
east, by the terrestrial paradise. It is a circle, in the centre
of which is represented the tree of the knowledge of good

FIG. 47.—FROM THE MAP IN HEREFORD CATHEDRAL.

and evil. Adam and Eve are there in company with the
serpent that beguiled them. The four legendary rivers come
out of the base of the tree, and they are seen below crossing
the map. Outside Eden the flight of the first couple, and the
angel that drove them away, are represented. At this extreme
eastern portion is the region of giants with the heads of

Essedons.

Tower of Babel.

Dragons

Pigmies.

The Monoceros.

The Mantichore.

A Sphinx.   The King of the Cyclops.   Blemmye.   Parasol lip.   Monocle

FIG. 48.—FROM THE MAP IN HEREFORD CATHEDRAL.

beasts. There, too, is seen the first human habitation, or
town, built by Enoch. Below appears the Tower of Babel.
Near this are two men seated on a hill close to the river
Jaxartes; one of them is eating a human leg and the other
an arm, which the legend explains thus:—"Here live the
Essedons, whose custom it is to sing at the funerals of their
parents; they tear the corpses with their teeth, and prepare
their food with these fragments of flesh, mixed with that of
animals. In their opinion it is more honourable to the dead
to be enclosed in the bodies of their relations than in those
of worms."

Below are seen dragons and pigmies, always to the east of
Asia, and a little further away in the midst of a strange
country, *the King of the Cyclops.*

This extraordinary geography shows us in India the "Mantichore, who has a triple range of teeth, the face of a man,
blue eyes, the red colour of blood, the body of a lion, and
the tail of a scorpion; its voice is a whistle."

On the north of the Ganges is represented a man with one
leg, shading his head with his foot, which is explained by the
following legend:—" In India dwell the Monocles, who have
only one leg, but who nevertheless move with surprising
velocity; when they wish to protect themselves from the heat
of the sun they make a shadow with the sole of their foot,
which is very large."

The Blemmys have their mouth and eyes in their chest;

others have their mouths and eyes on their shoulders. The Parvini are Ethiopians that have four eyes.

To the east of Syene is a man seated who is covering his head with his lip, "people who with their prominent lip shade their faces from the sun."

Above is drawn a little sun, with the word *sol*. Then comes an animal of human form, having the feet of a horse and the head and beak of a bird; he rests on a stick, and the legend tells us it is a satyr; the fauns, half men and half horses; the cynocephales—men with the head of a dog; and the cyanthropes—dogs with the heads of men. The sphinx has the wings of a bird, the tail of a serpent, the head of a woman. It is placed in the midst of the Cordilleras, which are joined to a great chain of mountains. Here lastly is seen the *monoceros*, a terrible animal; but here is the marvel: "When one shows to this *monoceros* a young girl, who, when the animal approaches, uncovers her breast, the monster, forgetting his ferocity, lays his head there, and when he is asleep may be taken defenceless."

Near to the lake Meotides is a man clothed in Oriental style, with a hat that terminates in a point, and holding by the bridle a horse whose harness is a human skin, and which is explained thus by the Latin legend: "Here live the griffins, very wicked men, for among other crimes they proceed so far as to make clothes for themselves and their horses out of the skins of their enemies."

More to the south is a large bird, the ostrich; according to the legend, "the ostrich has the head of a goose, the body of a crane, the feet of a calf; it eats iron."

Not far from the Riphean Mountains two men with long tunics and round bonnets are represented in the attitude of fighting; one brandishes a sword, the other a kind of club, and the legend tells us, "The customs of the people of the interior of Scythia are somewhat wild; they inhabit caves; they drink the blood of the slain by sucking their wounds; they pride themselves on the number of people they have slain—not to have slain any one in combat is reckoned disgraceful."

Near the river that empties itself into the Caspian Sea it is written: "This river comes from the infernal regions; it enters the sea after having descended from mountains covered with wood, and it is there, they say, that the mouth of hell opens."

To the south of this river, and to the north of Hyrcania, is represented a monster having the body of a man, the head, tail, and feet of a bull: this is the Minotaur. Further on are the mountains of Armenia, and the ark of Noah on one of its plateaux. Here, too, is seen a large tiger, and we read : "The tiger, when he sees that he has been deprived of his young, pursues the ravisher precipitately; but the latter, hastening away on his swift horse, throws a mirror to him and is safe."

U

Elsewhere appears Lot's wife changed into a pillar of salt ; the lynx who can see through a stone wall ; the river Lethe, so called because all who drink of it forget everything.

Numerous other details might be mentioned, but enough has been said to show the curious nature and exceeding interest of this map, in which matters of observation and imagination are strangely mixed. Another very curious geographical document of that epoch is the map of the world of the *Grandes Chroniques de Saint-Denis*. This belongs to the fourteenth century. The capitals here too are represented by edifices. The Mediterranean is a vertical canal, which goes from the Columns of Hercules to Jerusalem. The Caspian Sea communicates with it to the north, and the Red Sea to the south-east, by the Nile. It preserves the same position for Paradise and for the land of Gog and Magog that we have seen before. The geography of Europe is very defective. Britannia and Anglia figure as two separate islands, being represented off the west coast of Spain, with Allemania and Germania, also two distinct countries, to the north. The ocean is represented as round the whole, and the various points of the compass are represented by different kinds of winds on the outside.

This was the general style of the maps of the world at that period, as we may perceive from the various illustrations we have been able to give, and it curiously initiates us into the mediaeval ideas. Sometimes they are surrounded by laughable

figures of the winds with inflated cheeks, sometimes there
are drawn light children of Eolus seated on leathern bottles,
rotating the liquid within ; at other times, saints, angels,

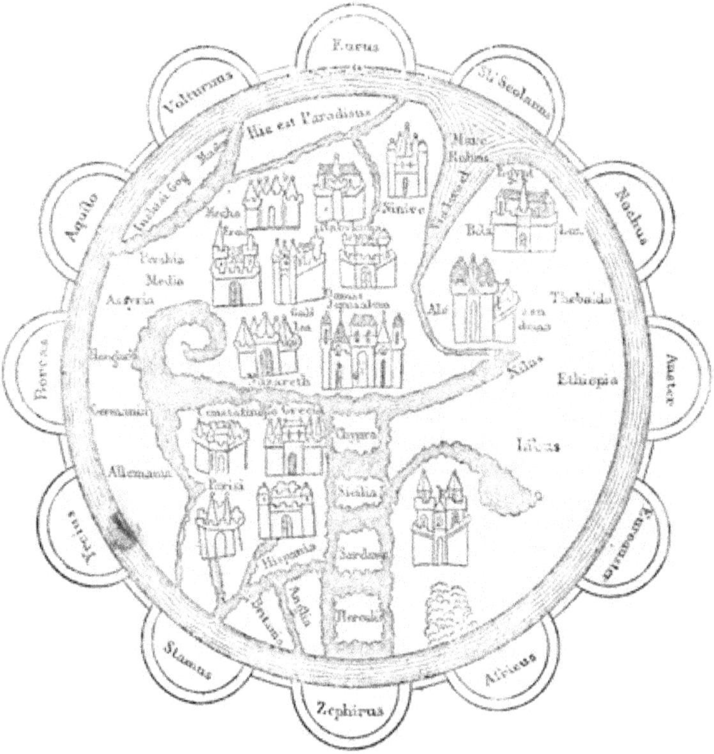

FIG. 49.—COSMOGRAPHY OF ST. DENIS.

Adam and Eve, or other people, adorn the circumference of
the map. Within are shown a profusion of animals, trees,
populations, monuments, tents, draperies, and monarchs

U 2

seated on their thrones—an idea which was useful, no
doubt, and which gave the reader some knowledge of the
local riches, the ethnography, the local forms of government
and of architecture in the various countries represented; but
the drawings were for the most part childish, and more
fantastic than real. The language, too, in which they were
written was as mixed as the drawings; no regularity was
preserved in the orthography of a name, which on the same
map may be written in ten different ways, being expressed in
barbarous Latin, Roman, or Old French, Catalan, Italian,
Castilian, or Portuguese!

During the same epoch other forms of maps in less detail
and of smaller size show the characters that we have seen in
the maps of earlier centuries.

Marco Polo, the traveller, at the end of the fourteenth
century, has preserved in his writings all the ancient traditions, and united them in a singular manner with the results
of his own observations. He had not seen Paradise, but he
had seen the ark of Noah resting on the top of Ararat. His
map of the world, preserved in the library at Stockholm, is
oval, and represents two continents.

In that which we inhabit, the only seas indicated are the
Mediterranean and the Black Sea. Asia appears at the east,
Europe to the north, and Africa to the south. The other
continent to the south of the equator, which is not marked,
is Antichthonia.

In a map of the world engraved on a medal of the fifteenth century during the reign of Charles V. there is still a reminiscence of the ideas of the concealed earth and Meropides, as described by Theopompus. We see the winds as cherubim ;

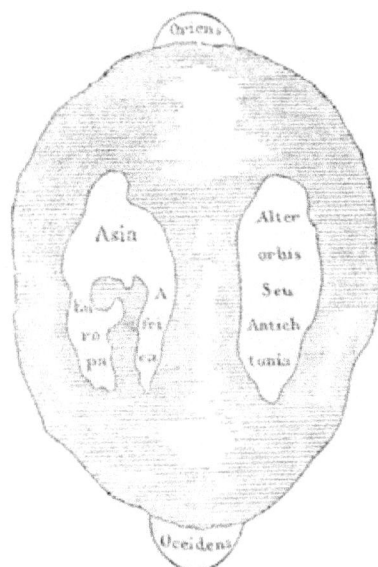

Fig. 50 —The Map of Marco Polo.

Europe more accurately represented than usual; but Africa still unknown, and a second continent, called Brumæ, instead of Antichthonia, with imaginary details upon it.

If such were the ideas entertained amongst the most enlightened nations, what may we expect among those who were less advanced ?   It would take us too long to describe

all that more Eastern nations have done upon this point since
the commencement of our present era, but we may give an
example or two from the Arabians.

In the ancient Arabian chronicle of Tabari is a system
founded on the earth being the solid foundation of all things;
we read: "The prophet says, the all-powerful and inimit-

FIG. 51.—MAP ON A MEDAL OF CHARLES V.

able Deity has created the mountain of Kaf round about the
earth; it has been called the foundation pile of the earth, as it
is said in the Koran, 'The mountains are the piles.' This
world is in the midst of the mountain of Kaf, just as the
finger is in the midst of the ring. This mountain is emerald,
and blue in colour; no man can go to it, because he would
have to pass four months in darkness to do so. There is in

that mountain neither sun, nor moon, nor stars ; it is so blue that the azure colour you see in the heavens comes from the brilliancy of the mountain of Kaf, which is reflected in the sky. If this were not so the sky would not be blue. All the mountains that you see are supported by Kaf; if it did not exist, all the earth would be in a continual tremble, and not a creature could live upon its surface. The heavens rest upon it like a tent."

Another Arabian author, Benakaty, writing in 1317, says : " Know that the earth has the form of a globe suspended in the centre of the heavens. It is divided by the two great circles of the meridian and equator, which cut each other at right angles, into four equal parts, namely, those of the north-west, north-east, south-west, and south-east. The inhabited portion of the earth is situated in the southern hemisphere, of which one half is inhabited."

Ibn-Wardy, who lived in the same century, adopted the idea of the ocean surrounding all the earth, and said we knew neither its depth nor its extent.

This ocean was also acknowledged by the author of the Kaf mountain ; he says it lies between the earth and that mountain, and calls it Bahr-al-Mohith.

The end of the fifteenth century saw the dawn of a new era in knowledge and science. The discoveries of Columbus changed entirely the aspect of matters, the imagination was excited to fresh enterprises, and the hardihood of the adventurers through good or bad success was such as want of liberty could not destroy.

Nevertheless, as we have seen, Columbus imagined the earth to have the shape of a pear. Not that he obtained this idea from his own observations, but rather retained it as a relic of past traditions. It is probable that it really dates from the seventh century. We may read in several cosmographical manuscripts of that epoch, that the earth has the form of a cone or a top, its surface rising from south to north. These ideas were considerably spread by the compilations of John of Beauvais in 1479, from whom probably Columbus derived his notion.

Although Columbus is generally and rightly known as the discoverer of the New World, a very curious suit was brought by Pinzon against his heirs in 1514. Pinzon pretended that the discovery was due to him alone, as Columbus had only followed his advice in making it. Pinzon told the admiral himself that the required route was intimated by an inspiration, or revelation. The truth was that this "revelation" was due to a flock of parrots, flying in the evening towards the south-west, which Pinzon concluded must be going in the direction of an invisible coast to pass the night in the bushes. Certainly the consequences of Columbus resisting the advice of Pinzon would have been most remarkable ; for had he continued to sail due west he would have been caught by the Gulf Stream and carried to Florida, or possibly to Virginia, and in this case the United States would have received a Spanish and Catholic population, instead of an English and Protestant one.

The discoveries of those days were often commemorated by the formation of heraldic devices for the authors of them, and we have in this way some curious coats of arms on record. That, for instance, of Sebastian Cano was a globe, with the legend, *Primus circumdedisti me.* The arms given to Columbus in 1493 consisted of the first map of America, with a range of islands in a gulf. Charles V. gave to Diego of Ordaz the figure of the Peak of Orizaba as his arms, to commemorate his having ascended it; and to the historian Oviedo, who passed thirty-four years without interruption (1513—47) in tropical America, the four beautiful stars of the Southern Cross.

We have arrived at the close of our history of the attempts that preceded the actual discovery of the form and constitution of the globe; since these were established our further progress has been in matters of detail. There now remains briefly to notice the attempts at discovering the size of the earth on the supposition, and afterwards certainty, of its being a globe.

The earliest attempt at this was made by Eratosthenes, 246 years before our era, and it was founded on the following reasoning. The sun illuminates the bottom of pits at Syene at the summer solstice; on the same day, instead of being

vertical over the heads of the inhabitants of Alexandria, it is
7¼ degrees from the zenith. Seven-and-a-quarter degrees
is the fiftieth part of an entire circumference; and the
distance between the two towns is five thousand stadia; hence
the circumference of the earth is fifty times this distance,
or 250 thousand stadia.

A century before our era Posidonius arrived at an analogous
result by remarking that the star Canopus touched the horizon
at Rhodes when it was 7 degrees 12 minutes above that of
Alexandria.

These measurements, which, though rough, were ingenious, were followed in the eighth century by similar
ones by the Arabian Caliph, Almamoun, who did not
greatly modify them.

The first men who actually went round the world were
the crew of the ship under Magellan, who started to the west in
1520; he was slain by the Philippine islanders in 1521, but his
ship, under his lieutenant, Sebastian Cano, returned by the
east in 1522. The first attempt at the actual measurement of
a part of the earth's surface along the meridian was made by
Fernel in 1528. His process was a singular, but simple one,
namely, by counting the number of the turns made by the
wheels of his carriage between Paris and Amiens. He made
the number 57,020, and accurate measurements of the distance many years after showed he had not made an error of
more than four turns.

The astronomer Picard attempted it again under Louis XIV. by triangulation.

The French astronomers have always been forward in this inquiry, and to them we owe the systematic attempts to arrive at a truer knowledge of the length of an arc of the meridian which were made in 1735-45 in Lapland and in Peru; and later under Mechain and Delambre, by order of the National Assembly, for the basis of the metrical system.

Observations of this kind have also been made by the English, as at Lough Foyle in Ireland, and in India.

The review which has here been made of the various ideas on what now seems so simple a matter cannot but impress us with the vast contrast there is between the wild attempts of the earlier philosophers and our modern affirmations. What progress has been made in the last two thousand years! And all of this is due to hard work. The true revelation of nature is that which we form ourselves, by our persevering efforts. We now know that the earth is approximately spherical, but flattened by about $\frac{1}{300}$ at the poles, is three-quarters covered with water, and enveloped everywhere by a light atmospheric mantle. The distance from the centre of the earth to its surface is 3,956 miles, its area is 197 million square miles, its volume is 256,000 millions of cubic miles, its weight is six thousand trillion tons. So, thanks to the bold measurements of its inhabitants, we know as much about it as we are likely to know for a long time to come.

CHAPTER XI.

THE legends that were for so many ages prevalent in Europe had their foundation in the attempt to make the accounts of Scripture and the ideas and dogmas of the Fathers of the Church fit into the few and insignificant facts that were known with respect to the earth, and the system of which it forms a part, and the far more numerous imaginations that were entertained about it.

We are therefore led on to examine some of these legends, that we may appreciate how far a knowledge of astronomy will effect the eradication of errors and fantasies which, under the aspect of truth, have so long enslaved the people. No doubt the authors of the legendary stories knew well enough their allegorical nature ; but those who received them supposed that they gave true indications of the nature of the earth and world, and therefore accepted them as facts.

Some indeed considered that the whole physical constitution of the world was a scaffold or a model, and that there was a real theological universe hidden beneath this semblance. No one omitted from his system the spiritual heaven in which the angels and just men might spend their existence; but in addition to this there were places whose reality was believed in, but whose locality is more difficult to settle, and which therefore were moved from one place to another by various writers, viz., the infernal regions, purgatory, and the terrestrial paradise.

We will here recount some of those legends, which wielded sufficient sway over men's minds as to gain their belief in the veritable existence of the places described, and in this way to influence their astronomical and cosmographical ideas.

And for the first we will descend to the infernal regions with Plutarch and Thespesius.

This Thespesius relates his adventures in the other world. Having fallen head-first from an elevated place, he found himself unwounded, but was contused in such a way as to be insensible. He was supposed to be dead, but, after three days, as they were about to bury him, he came to life again. In a few days he recovered his former powers of mind and body; but made a marvellous change for the better in his life.

He said that at the moment that he lost consciousness he found himself like a sailor at the bottom of the sea; but afterwards, having recovered himself a little, he was able to breathe perfectly, and seeing only with the eyes of his soul,

he looked round on all that was about him. He saw no longer
the accustomed sights, but stars of prodigious magnitude,
separated from each other by immense distances. They were
of dazzling brightness and splendid colour. His soul, carried
like a vessel on the luminous ocean, sailed along freely and
smoothly, and moved everywhere with rapidity. Passing
over in silence a large number of the sights that met his eye,
he stated that the souls of the dead, taking the form of
bubbles of fire, rise through the air, which opens a passage
above them; at last the bubbles, breaking without noise, let
out the souls in a human form and of a smaller size, and
moving in different ways. Some, rising with astonishing
lightness, mounted in a straight line; others, running round
like a whipping-top, went up and down by turns with a confused and irregular motion, making small advance by long
and painful efforts. Among this number he saw one of his
parents, whom he recognised with difficulty, as she had died
in his infancy; but she approached him, and said, "Good
day, Thespesius." Surprised to hear himself called by this
name, he told her that he was called Arideus, and not
Thespesius. "That was once your name," she replied,
"but in future you will bear that of Thespesius, for you
are not dead, only the intelligent part of your soul has
come here by the particular will of the gods; your other
faculties are still united to your body, which keeps them
like an anchor. The proof I will give you is that the souls

of the dead do not cast any shadow, and they cannot move their eyes."

Further on, in traversing a luminous region, he heard, as he was passing, the shrill voice of a female speaking in verse, who presided over the time Thespesius should die. His genius told him that it was the voice of the Sibyl, who, turning on the orbit of the moon, foretold the future. Thespesius would willingly have heard more, but, driven off by a rapid whirlwind, he could make out but little of her predictions. In another place he remarked several parallel lakes, one filled with melted and boiling gold, another with lead colder than ice, and a third with very rough iron. They were kept by genii, who, armed with tongs like those used in forges, plunged into these lakes, and then withdrew by turns, the souls of those whom avarice or an insatiable cupidity had led into crime; after they had been plunged into the lake of gold, where the fire made them red and transparent, they were thrown into the lake of lead. Then, frozen by the cold, and made as hard as hail, they were put into the lake of iron, where they became horribly black. Broken and bruised on account of their hardness, they changed their form, and passed once more into the lake of gold, and suffered in these changes inexpressible pain.

In another place he saw the souls of those who had to return to life and be violently forced to take the form of all sorts of animals. Among the number he saw the soul of

Nero, which had already suffered many torments, and was bound with red-hot chains of iron. The workmen were seizing him to give him the form of a viper, under which he was destined to live, after having devoured the womb that bore him.

The locality of these infernal regions was never exactly determined. The ancients were divided upon the point. In the poems of Homer the infernal regions appear under two different forms: thus, in the *Iliad*, it is a vast subterranean cavity; while in the *Odyssey*, it is a distant and mysterious country at the extremity of the earth, beyond the ocean, in the neighbourhood of the Cimmerians.

The description which Homer gives of the infernal region proves that in his time the Greeks imagined it to be a copy of the terrestrial world, but one which had a special character. According to the philosophers it was equally remote from all parts of the earth. Thus Cicero, in order to show that it was of no consequence where one died, said, wherever we die there is just as long a journey to be made to reach the " infernal regions."

The poets fixed upon certain localities as the entrance to this dismal empire: such was the river Lethe, on the borders of the Scythians; the cavern Acherusia in Epirus, the mouth of Pluto, in Laodicœa, the cave of Zenarus near Lacedæmon.

In the map of the world in the *Polychronicon* of Ranulphus Uygden, now in the British Museum, it is stated: " The Island of Sicily was once a part of Italy. There is Mount

Etna, containing the infernal regions and purgatory, and it
has Scylla and Charybdis, two whirlpools."

Ulysses was said to reach the place of the dead by crossing
the ocean to the Cimmerian land, Æneas to have entered it
by the Lake of Avernus. Xenophon says that Hercules went
there by the peninsula of Arechusiade.

Much of this, no doubt, depends on the exaggeration and
misinterpretation of the accounts of voyagers; as when the
Phœnicians related that, after passing the Columns of Hercules,
to seek tin in Thule and amber in the Baltic, they came,
at the extremity of the world, to the Fortunate Isles, the
abode of eternal spring, and further on to the Hyperborean
regions, where a perpetual night enveloped the country—the
imagination of the people developed from this the Elysian
fields, as the places of delight in the lower regions, having
their own sun, moon, and stars, and Tartarus, a place of
shades and desolation.

In every case, however, both among pagans and Christians,
the locality was somewhere in the centre of the earth. The
poets and philosophers of Greece and Rome made very
detailed and circumstantial maps of the subterranean regions.
They enumerated its rivers, its lakes, and woods, and mountains, and the places where the Furies perpetually tormented
the wicked souls who were condemned to eternal punishment.
These ideas passed naturally into the creeds of Christians
through the sect of the Essenes, of whom Josephus writes as

follows:—" They thought that the souls of the just go beyond
the ocean to a place of repose and delight, where they were
troubled by no inconvenience, no change of seasons. Those
of the wicked, on the contrary, were relegated to places
exposed to all the inclemencies of the weather, and suffered
eternal torments. The Essenes," adds the same author, "have
similar ideas about these torments to those of the Greeks
about Tartarus and the kingdom of Pluto. The greater part
of the Gnostic sects, on the contrary, considered the lower
regions as simply a place of purgatory, where the soul is
purified by fire."

Amongst all the writings of Christian ages in which matters
such as we are now passing in review are described, there is
one that stands out beyond all others as a masterpiece, and
that is the magnificent poem of Dante, his *Divine Comedy*,
wherein he described the infernal regions as they presented
themselves to his lively and fertile imagination. We have in
it a picture of mediæval ideas, painted for us in indelible
lines, before the remembrance of them was lost in the past.
The poem is at once a tomb and a cradle—the tomb of a
world that was passing, the cradle of the world that was to
come: a portico between two temples, that of the past and
that of the future. In it are deposited the traditions, the
ideas, the sciences of the past, as the Egyptians deposited
their kings and symbolic gods in the sepulchres of Thebes
and Memphis. The future brings into it its aspirations and

its germs enveloped in the swaddling clothes of a rising
language and a splendid poetry—a mysterious infant that is
nourished by the two teats of sacred tradition and profane
fiction, Moses and St. Paul, Homer and Virgil.

The theology of Dante, strictly orthodox, was that of St.
Thomas and the other doctors of the Church. Natural philosophy, properly so called, was not yet in existence. In
astronomy, Ptolemy reigned supreme, and in the explanation
of celestial phenomena no one dreamt or dared to dream of
departing in any way from the traditionally sacred system.

In those days astronomy was indissolubly linked with a
complete series of philosophical and theological ideas, and
included the physics of the world, the science of life in
every being, of their organisation, and the causes on which
depended the aptitudes, inclinations, and even in part the
actions, of men, the destinies of individuals, and the events
of history. In this theological, astronomical, and terrestrial
universe everything emanated from God; He had created
everything, and the creation embraced two orders of beings,
the immaterial and the corporeal.

The pure spirits composed the nine choirs of the celestial
hierarchy. Like so many circles, they were ranged round a
fixed point, the Eternal Being, in an order determined by their
relative perfection. First the seraphim, then the cherubim,
and afterwards the simple angels. Those of the first circle
received immediately from the central point the light and

the virtue which they communicated to those of the second ; and so on from circle to circle, like mirrors which reflect, with an ever-lessening light, the brilliancy of a single luminous point. The nine choirs, supported by Love, turned without ceasing round their centre in larger and larger circles according to their distance ; and it was by their means that the motion and the divine inflatus was communicated to the material creation.

This latter had in the upper part of it the empyreal, or heaven of pure light. Below that, was the *Primum mobile*, the greatest body in the heavens, as Dante calls it, because it surrounds all the rest of the circle, and bounds the material world. Then came the heaven of the fixed stars ; then, continuing to descend, the heavens of Saturn, Jupiter, Mars, the sun, Venus, Mercury, the moon, and lastly, the earth, whose solid and compact nucleus is surrounded by the spheres of water, air, and fire.

As the choirs of angels turn about a fixed point, so the nine material circles turn also about another fixed point, and are moved by the pure spirits.

Let us now descend to the geography of the interior of the earth. Within the earth is a large cone, whose layers are the frightful abodes of the condemned, and which ends in the centre, where the divine Justice keeps bound up to his chest in ice the prince of the rebellious angels, the emperor of the kingdom of woe. Such are the infernal regions which

Dante describes according to ideas generally admitted in the middle ages.

The form of the infernal regions was that of a funnel or reversed cone. All its circles were concentric, and continually diminished ; the principal ones were nine in number. Virgil also admitted nine divisions—three times three, a number sacred *par excellence.* The seventh, eighth, and ninth circles were divided into several regions ; and the space between the entrance to the infernal regions and the river Acheron, where the resting-place of the damned really commenced, was divided into two parts. Dante, guided by Virgil, traversed all these circles.

It was in 1300 that the poet, " in the midst of the course of life," at the age of thirty-five, passed in spirit through the three regions of the dead. Lost in a lonely, wild, and dismal forest, he reached the base of a hill, which he attempted to climb. But three animals, a panther, a lion, and a thin and famished wolf, prevented his passage ; so, returning again where the sun was powerless, into the shades of the depths of the valley, there met him a shadow of the dead. This human form, whom a long silence had deprived of speech, was Virgil, who was sent to guide and succour him by a celestial dame, Beatrice, the object of his love, who was at the same time a real and a mystically ideal being.

Virgil and Dante arrived at the gate of the infernal regions ; they read the terrible inscription placed over the gate; they entered and found first those unhappy souls who had lived without virtue and without vice. They reached the banks of Acheron and saw Charon, who carried over the souls in his bark to the other side; and Dante was surprised by a profound sleep. He woke beyond the river, and he descended into the Limbo which is the first circle of the infernal regions. He found there the souls of those who had died without baptism, or who had been indifferent to religion.

They descended next to the second circle, where Minos, the judge of those below, is enthroned. Here the luxurious are punished. The poet here met with Francesca of Rimini and Paul, her friend. He completely recovered the use of his senses, and passed through the third circle, where the gourmands are punished. In the fourth he found Plutus, who guards it. Here are tormented the prodigal and the avaricious. In the fifth are punished those who yield to anger. Dante and Virgil there saw a bark approaching, conducted by Phlegias; they entered it, crossed a river, and arrived thus at the base of the red-hot iron walls of the infernal town of Dite. The demons that guarded the gates refused them admittance, but an angel opened them, and the two travellers there saw the heretics that were enclosed in tombs surrounded by flames.

The travellers then visited the circles of violence, fraud, and usury, when they came to a river of blood guarded by a troop of centaurs; suddenly they saw coming to them

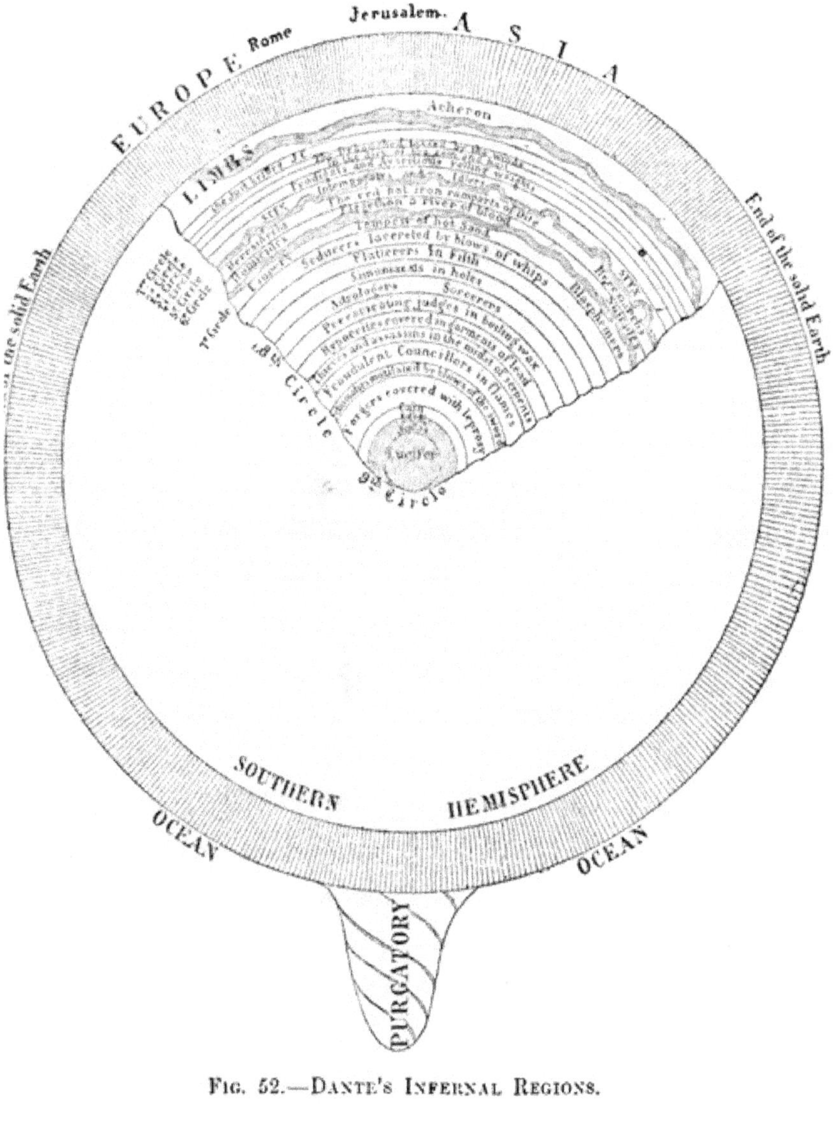

FIG. 52.—DANTE'S INFERNAL REGIONS.

Geryon, who represents fraud, and this beast took them behind him to carry them across the rest of the infernal space.

The eighth circle was divided into ten valleys, comprising: the flatterers; the simoniacal; the astrologers; the sorcerers; the false judges; the hypocrites who walked about clothed with heavy leaden garments; the thieves, eternally stung by venomous serpents; the heresiarchs; the charlatans, and the forgers.

At last the poets descended into the ninth circle, divided into four regions, where are punished four kinds of traitors. Here is recounted the admirable episode of Count Ugolin. In the last region, called the region of Judas, LUCIFER is enchained. There is the centre of the earth, and Dante, hearing the noise of a little brook, reascended to the other hemisphere, on the surface of which he found, surrounded by the Southern Ocean, the mountain of Purgatory.

Such was the famous *Inferno* of Dante.

Not only was the geography of the infernal regions attempted in the middle ages, but even their size. Dexelius calculated that the number of the damned was a hundred millions, and that their abode need not measure more than one German mile in every direction. Cyrano of Bergerac amusingly said that it was the damned that kept turning the earth, by hanging on the ceiling like bats, and trying to get away.

In 1757 an English clergyman, Dr. Swinden, published
a book entitled, *Researches on the Nature of the Fire of
Hell and the Place where it is situated.* He places it in
the sun. According to him the Christians of the first
century had placed it beneath the earth on account of a
false interpretation of the descent of Jesus into hell after
his crucifixion, and by false ideas of cosmography. He
attempted to show, 1st, that the terrestrial globe is too
small to contain even the angels that fell from heaven after
their battle; 2nd, that the fire of hell is real, and that the
closed globe of earth could not support it a sufficiently long
period; 3rd, that the sun alone presents itself as the necessary place, being a well-sustained fire, and directly opposite
in situation to heaven, since the empyreal is round the
outside of the universe, and the sun in the centre. What
a change to the present ideas, even of doctors of divinity,
in a hundred years !

So far, then, for mediæval ideas on the position and
character of hell. Next as to purgatory.

The voyage to purgatory that has met with most success
is certainly the celebrated Irish legend of St. Patrick,
which for several centuries was admitted as authentic, and
the account of which was composed certainly a century
before the poem of Dante.

This purgatory, the entrance to which is drawn in more
than one illuminated manuscript, is situated in Ireland, on

one of the islands of Lough Derg, County Donegal, where there are still two chapels and a shrine, at which annual ceremonies are performed. A knight, called Owen, resolved to visit it for penance; and the chronicle gives us an account of his adventures.

First he had his obsequial rites performed, as if he had been dead, and then he advanced boldly into the deep ravine; he marched on courageously, and entered into the semi-shadows; he marched on, and even this funereal twilight abandoned him, and "when he had gone for a long time in this obscurity, there appeared to him a little light as it were from a glimmer of day." He arrived at a house, built with much care, an imposing mansion of grief and hope, a marvellous edifice, but similar nevertheless to a monkish cloister, where there was no more light than there is in this world in winter at vesper-time.

The knight was in dreadful suspense. Suddenly he heard a terrible noise, as if the universe was in a riot; for it seemed certainly to him as if every kind of beast and every man in the world were together, and each gave utterance to their own cry, at one time and with one voice, so that they could not make a more frightful noise.

Then commenced his trials, and discourse with the infernal beings; the demons yelled with delight or with fury round him. "Miserable wretch," said some, "you are come here to suffer." " Fly," said others, "for you have not

PLATE XII.—THE LEGEND OF OWEN.

behaved well in the time that is passed : if you will take our advice, and will go back again to the world, we will take it as a great favour and courtesy."

Owen was thrown on the dark shadowy earth, where the demons creep like hideous serpents. A mysterious wind, which he scarcely heard, passed over the mud, and it seemed to the knight as if he had been pierced by a spear-head. After a while the demons lifted him up ; they took him straight off to the east, where the sun rises, as if they were going to the place where the universe ends. " Now, after they had journeyed for a long time here and there over divers countries, they brought him to an open field, very long and very full of griefs and chastisements ; he could not see the end of the field, it was so long ; there were men and women of various ages, who lay down all naked on the ground with their bellies downwards, who had hot nails driven into their hands and feet ; and there was a fiery dragon, who sat upon them and drove his teeth into their flesh, and seemed as if he would eat them ; hence they suffered great agony, and bit the earth in spite of its hardness, and from time to time they cried most piteously ' Mercy, mercy ;' but there was no one there who had pity or mercy, for the devils ran among them and over them, and beat them most cruelly."

The devils brought the knight towards a house of punishment, so broad and long that one could not see the end.

This house is the house of baths, like those of the infernal regions, and the souls that are bathed in ignominy are there heaped in large vats. "Now so it was, that each of these vats was filled with some kind of metal, hot and boiling, and there they plunged and bathed many people of various ages, some of whom were plunged in over their heads, others up to the eyebrows, others up to the eyes, and others up to the mouth. Now all in truth of these people cried out with a loud voice and wept most piteously."

Scarcely had the knight passed this terrible place, and left behind in his mysterious voyage that column of fire which rose like a lighthouse in the shades, and which shone so sadly betwixt hope. and eternal despair, than a vast and magnificent spectacle displayed itself in the subterranean space.

This luminous and odorescent region, where one might see so many archbishops, bishops, and monks of every order, was the terrestrial paradise; man does not stay there always; they told the knight that he could not taste too long its rapid delights; it is a place of transition between purgatory and the abodes of heaven, just as the dark places which he had traversed were made by the Creator between the world and the infernal regions.

"In spite of our joys," said the souls, "we shall pass away from here." Then they took him to a mountain, and told him to look, and asked of him what colour the heavens seemed to be there where he was standing, and he replied it was the colour of burning gold, such as is in the furnace; and then they said to him, "That which you see is the entrance to heaven and the gate of paradise."

The attempts at identification of hell and purgatory have not been so numerous, perhaps because the subjects were not very attractive, except as the spite of men might think of them in reference to other people; but when we come to the terrestrial paradise, quite a crowd of attempts by every kind of writer to fix its position in any and every part of the globe is met with on every side.

In the seventeenth century, under Louis XIV., Daniel Huet, Bishop of Avranches, gave great attention to the question, and collected every opinion that had been expressed upon it, with a view to arriving at some definite conclusion for himself. He was astonished at the number of writings and the diversity of the opinions they expressed.

"Nothing," he says, "could show me better how little is really known about the situation of the terrestrial paradise than the differences in the opinions of those who have occupied themselves about the question. Some have placed it in the third heaven, some in the fourth, in the heaven of the moon, in the moon itself, on a mountain near the lunar heaven, in the middle region of the air, out of the earth,

upon the earth, beneath the earth, in a place that is hidden
and separated from man.   It has been placed under the
North Pole, in Tartary, or in the place now occupied by the
Caspian Sea.   Others placed it in the extreme south, in the
land of fire.   Others in the Levant, or on the borders of the
Ganges, or in the Island of Ceylon, making the name India
to be derived from Eden, the land where the paradise was
situated.   It has been placed in China, or in an inaccessible
place beyond the Black Sea; by others in America, in Africa,
beneath the equator, in the East, &c. &c."

Notwithstanding this formidable array, the good bishop
was bold enough to make his choice between them all.   His
opinion was that the dwelling-place of the first man was
situated between the Tigris and Euphrates, above the place
where they separate before falling into the Persian Gulf;
and, founding this opinion on very extensive reading, he
declared that of all his predecessors, Calvin had come nearest
to the truth.

Among the other authors of greater or less celebrity that
have occupied themselves in this question, we may instance
the following :—

Raban Maur (ninth century) believed that the terrestrial
paradise was at the eastern extremity of the earth.   He
described the tree of life, and added that there was
neither heat nor cold in that garden ; that immense rivers of
water nourished all the forest; and that the paradise was

surrounded by a wall of fire, and its four rivers watered the earth.

James of Vitry supposed Pison to come out of the terrestrial paradise. He describes also the garden of Eden ; and, like all the cosmographers of the middle ages, he placed it in the most easterly portion of the world in an inaccessible place, and surrounded by a wall of fire, which rose up to heaven.

Dati placed also the terrestrial paradise in Asia, like the cosmographers that preceded him, and made the Nile come from the east. Stenchus, the librarian of St. Siége, who lived in the sixteenth century, devoted several years to the problem, but discovered nothing. The celebrated orientalist and missionary Bochart wrote a treatise on this subject in 1650. Thévenot published also in the seventeenth century a map representing the country of the Lybians, and adds that " several great doctors place the terrestrial paradise there."

An Armenian writer who translated and borrowed from St. Epiphanius (eighth century) produced a *Memorial on the Four Rivers of the Terrestrial Paradise.* He supposes they rise in the unknown land of the Amazons, whence also arise the Danube and the Hellespont, and they deliver their waters into that great sea that is the source of all seas, and which surrounds the four quarters of the globe. He afterwards says, following up the same theory, that the rivers of paradise surround the world and enter again into the sea, which is the universal ocean."

Gervais and Robert of St. Marien d'Auxerre taught that the terrestrial paradise was on the eastern border of the *squa.e* which formed the world. Alain de Lille, who lived in the thirteenth century, maintained in his *Anticlaudianus* that the earth is circular, and the garden of Eden is in the east of Asia. Joinville, the friend of St. Louis, gives us a curious notion of his geographical ideas, since, with regard to paradise, he assures us that the four great rivers of the south come out of it, as do the spices. "Here," he says, referring to the Nile, "it is advisable to speak of the river which passes by the countries of Egypt, and comes from the terrestrial paradise. Where this river enters Egypt there are people very expert and experienced, as thieves are here, at stealing from the river, who in the evening throw their nets on the streams and rivers, and in the morning they often find and carry off the spices which are sold here in Europe as coming from Egypt at a good rate, and by weight, such as cinnamon, ginger, rhubarb, cloves, lignum, aloes, and several other good things, and they say that these good things come *from the terrestrial paradise*, and that the wind blows them off the trees that are growing there." And he says that near the end of the world are the peoples of Gog and Magog, who will come at the end of the world with Antichrist.

We find, however, more than descriptions—we have representations of the terrestrial paradise by cartographers of the middle ages, some of which we have seen in speaking of their general ideas of geography, and we will now introduce others.

Fra Mauro, a religious cosmographer of the fifteenth century, gives on the east side of a map of the world a repre-

FIG. 53.—PARADISE OF FRA MAURO.

sentation which shows us that at that epoch the "garden of delights" had become very barren. It is a vast plain, on which we see Jehovah and the first human couple, with a circular rampart surrounding it. The four rivers flow out of it by bifurcating. An angel protects the principal gate, which cannot be reached but by crossing barren mountains.

The cosmographical map of Gervais, dedicated to the Emperor Otho IV., shows the terrestrial paradise in the centre of the earth, which is square, and is situated in the midst of the seas. Adam and Eve appear in consultation.

The map of the world prepared by Andreas Bianco, in the fifteenth century, represents Eden, Adam and Eve, and the tree of life. On the left, on a peninsula, are seen the reprobated people of Gog and Magog, who are to accompany Antichrist. Alexander is also represented there, but without apparent reason. The paradisaical peninsula has a building on it with this inscription, " Ospitius Macarii."

Formalconi says, on this subject, that a certain Macarius lives near paradise, who is a witness to all that the author states, and as Bianco has indicated, his cell was close to the gates of paradise.

This legend has reference to the pilgrims of St. Macarius, a tradition that was spread on the return of the Crusaders, of three monks who undertook a voyage to discover the point where the earth and heaven meet, that is 'to say, the place of the terrestrial paradise. The map of Rudimentum, a vast compilation published at Lübeck in 1475 by the Dominican Brocard, represents the terrestrial paradise surrounded by walls, but it is less sterile that in the last picture, as may be seen on the next page.

In the year 1503, when Varthema, the adventurous Bolognian, went to the Indies by the route of Palestine and Syria,

he was shown the evil-reputed house which Cain dwelt in,
which was not far from the terrestrial paradise. Master
Gilius, the learned naturalist who travelled at the expense
of Francis I., had the same satisfaction. The simple faith

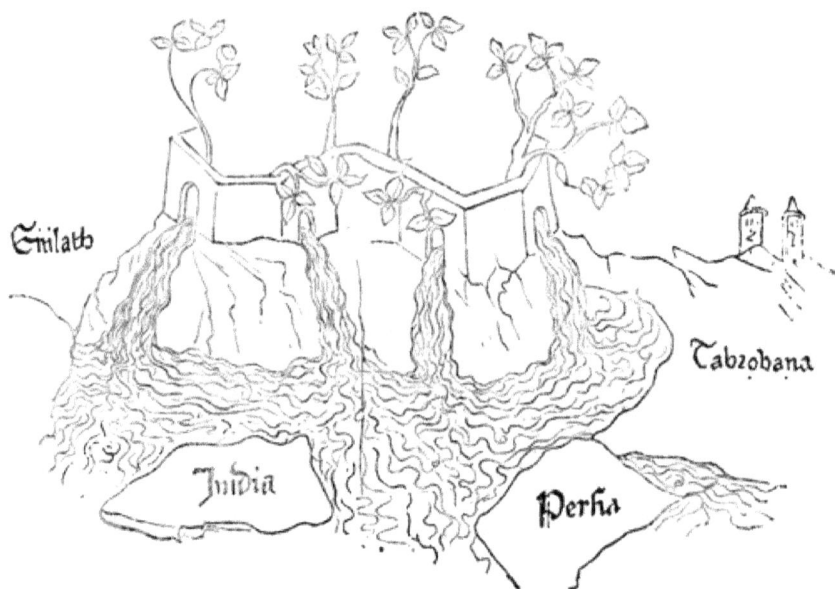

FIG. 54.—THE PARADISE OF THE FIFTEENTH CENTURY.

of our ancestors had no hesitation in accepting such
archæology.

The most curious and interesting of all attempts to discover
the situation of paradise was that made half unconsciously
by Columbus when he first found the American shore.

In his third voyage, when for the first time he reached the

main land, he was persuaded not only that he had arrived at
the extremity of Asia, but that he could not be far from the
position of paradise. The Orinoco seemed to be one of those
four great rivers which, according to tradition, came out of
the garden inhabited by our first parents, and his hopes
were supported by the fragrant breezes that blew from the
beautiful forests on its banks. This, he thought, was but the
entrance to the celestial dwelling-place, and if he had dared—
if a religious fear had not held back him who had risked
everything amidst the elements and amongst men, he would
have liked to push forward to where he might hope to find
the celestial boundaries of the world, and, a little further, to
have bathed his eyes, with profound humility, in the light of
the flaming swords which were wielded by two seraphim
before the gate of Eden.

He thus expresses himself on this subject in his letter to
one of the monarchs of Spain, dated Hayti, October, 1498.
" The Holy Scriptures attest that the Lord created paradise,
and placed in it the tree of life, and made the four great rivers
of the earth to pass out of it, the Ganges of India, the
Tigris, the Euphrates (passing from the mountains to form
Mesopotamia, and ending in Persia), and the Nile, which
rises in Ethiopia and goes to the Sea of Alexander.   I
cannot, nor have been ever able to find in the books of the
Latins or Greeks anything authentic on the site of this
terrestrial paradise, nor do I see anything more certain in

the maps of the world. Some place it at the source of the
Nile, in Ethiopia; but the travellers who have passed
through those countries have not found either in the mildness of the climate or in the elevation of the site towards
heaven anything that could lead to the presumption that
paradise was there, and that the waters of the Deluge were
unable to reach it or cover it. Several pagans have written
for the purpose of proving it was in the Fortunate Isles, which
are the Canaries. St. Isidore, Bede, and Strabo, St. Ambrosius,
Scotus, and all judicious theologians affirm with one accord
that paradise was in the East. It is from thence only that
the enormous quantity of water can come, seeing that the
course of the rivers is extremely long; and these waters (of
paradise) arrive here, where I am, and form a lake. There
are great signs here of the neighbourhood of the terrestrial
paradise, for the site is entirely conformable to the opinion of
the saints and judicious theologians. The climate is of admirable mildness. I believe that if I passed beneath the equinoctial line, and arrived at the highest point of which I have
spoken, I should find a milder temperature, and a change in
the stars and the waters; not that I believe that the point
where the greatest height is situated is navigable, or even
that there is water there, or that one could reach it, but I
am convinced that *there* is the terrestrial paradise, where no
one can come except by the will of God."

In the opinion of this illustrious navigator the earth had

the form of a pear, and its surface kept rising towards the east, indicated by the point of the fruit. It was there that he supposed might be found the garden where ancient tradition imagined the creation of the first human couple was accomplished.

We can scarcely think without astonishment of the great amount of darkness that obscured scientific knowledge, when this great man appeared on the scene of the world, nor of the rapidity with which the obscurity and vagueness of ideas were dissipated almost immediately after his marvellous discoveries. Scarcely had a half century elapsed after his death, than all the geographical fables of the middle ages did no more than excite smiles of incredulity, although during his life the universal opinion was not much advanced upon the times of the famous knight John of Mandeville, who wrote gravely as follows :—

" No mortal man can go to or approach this paradise. By land no one can go there on account of savage beasts which are in the deserts, and because of mountains and rocks that cannot be passed over, and dark places without number ; nor can one go there any better by sea ; the water rushes so wildly, it comes in so great waves, that no vessel dare sail against them. The water is so rapid, and makes so great a noise and tempest, that no one can hear however loud he is spoken to, and so when some great men with good courage have attempted several times to go by this river to paradise, in large companies, they have never been able to accomplish their journey. On the contrary, many have died with fatigue in swimming against the watery waves. Many others have become blind, others have become deaf by the noise of the water, and others have been suffocated and lost in the waves, so that no mortal man can approach it except by the special grace of God."

With one notable exception, no attempts have been made of late years to solve such a question. That exception is by the noble and indefatigable Livingstone, who declared his conviction to Sir Roderick Murchison, in a letter published in the *Athenæum*, that paradise was situated somewhere near the sources of the Nile.

Those generally who now seek an answer to the question of the birthplace of the human race do not call it paradise.

Since man is here, and there was a time quite recent, geologically speaking, when he was not, there must have been some actual locality on the earth's surface where he was first a man. Whether we have, or even can hope to have, enough information to indicate where that locality was situated, is a matter of doubt. We have not at present. Those who have attended most to the subject appear to think some island the most probable locality, but it is quite conjectural.

The name " Paradise " appears to have been derived from the Persian, in which it means a garden; similarly derived words express the same idea in other languages; as in the

Hebrew *pardés*, in the Arabian *firdans*, in the Syriac *pardiso*, and in the Armenian *partes*. It has been thought that the Persian word itself is derived from the Sanscrit *pradesa*, or *paradesa*, which means a circle, a country, or strange region ; which, though near enough as to sound, does not quite agree as to meaning. " Eden " is from a Hebrew root meaning delights.

CHAPTER XII.

WE have seen in the earlier chapters on the systems of the ancients and their ideas of the world how everything was once supposed to have exclusive reference to man, and how he considered himself not only chief of animate objects, but that his own city was the centre of the material world, and his own world the centre of the material universe; that the sun was made to shine, as well as the moon and stars, for his benefit; and that, were it not for him they would have no reason for existence. And we have seen how, step by step, these illusions have been dispelled, and he has learnt to appreciate his own littleness in proportion as he has realised the immensity of the universe of which he forms part.

If such has been his history, and such his former ideas on the regular parts, as we may call them, of nature, much more have similar ideas been developed in relation to those other phenomena which, coming at such long intervals, have not been recognised by him as periodic, but have seemed to have some relation to mundane affairs, often of the smallest consequence. Such are eclipses of the sun and moon, comets, shooting-stars, and meteors. Among the less instructed of men, even when astronomers of the same age and nation knew their real nature, eclipses have always been looked upon as something ominous of evil.

Among the ancient nations people used to come to the assistance of the moon, by making a confused noise with all kinds of instruments, when it was eclipsed. It is even done now in Persia and some parts of China, where they fancy that the moon is fighting with a great dragon, and they think the noise will make him loose his hold and take to flight. Among the East Indians they have the same belief that when the sun and the moon are eclipsed, a dragon is seizing them, and astronomers who go there to observe eclipses are troubled by the fears of their native attendants, and by their endeavours to get into the water as the best place under the circumstances. In America the idea is that the sun and moon are tired when they are eclipsed. But the more refined Greeks believed for a long time that the moon was bewitched, and that the magicians made it descend from heaven, to put into the herbs a certain maleficent froth. Perhaps the idea of the Dragon arose from the ancient custom of calling the places in the heavens at which the eclipses of the moon took place the head and tail of the Dragon.

In ancient history we have many curious instances of the
very critical influence that eclipses have had, especially in
the case of events in a campaign, where it was thought
unfavourable to some projected attempt.

Thus an eclipse of the moon was the original cause of the
death of the Athenian general Nicias.   Just at a critical
juncture, when he was about to depart from the harbour
of Syracuse, the eclipse filled him and his whole army
with dismay.   The result of his terror was that he delayed
the departure of his fleet, and the Athenian army was
cut in pieces and destroyed, and Nicias lost his liberty
and life.

Plutarch says they could understand well enough the
cause of the eclipse of the sun by the interposition of the
moon, but they could not imagine by the opposition of what
body the moon itself could be eclipsed.

One of the most famous eclipses of antiquity was that of
Thales, recorded by Herodotus, who says:—"The Lydians
and the Medes were at war for five consecutive years.   Now
while the war was sustained on both sides with equal chance,
in the sixth year, one day when the armies were in battle
array, it happened that in the midst of the combat the day
suddenly changed into night.   Thales of Miletus had predicted this phenomenon to the Ionians, and had pointed out
precisely that very year as the one in which it would take
place.   The Lydians and Medes, seeing the night succeeding

suddenly to the day, put an end to the combat, and only cared to establish peace."

Another notable eclipse is that related by Diodorus Siculus. It was a total eclipse of the sun, which took place while Agathocles, fleeing from the port of Syracuse, where he was blockaded by the Carthaginians, was hastening to gain the coast of Africa. "When Agathocles was already surrounded by the enemy, night came on, and he escaped contrary to all hope. On the day following so complete an eclipse of the sun took place that it seemed altogether night, for the stars shone out in all places. The soldiers therefore of Agathocles, persuaded that the gods were intending them some misfortune, were in the greatest perturbation about the future. Agathocles was equal to the occasion. When disembarked in Africa, where, in spite of all his fine words, he was unable to reassure his soldiers, whom the eclipse of the sun had frightened, he changed his tactics, and pretending to understand the prodigy, "I grant, comrades," he said, "that had we perceived this eclipse before our embarkation we should indeed have been in a critical situation, but now that we have seen it after our departure, and as it always signifies a change in the present state of affairs, it follows that our circumstances, which were very bad in Sicily, are about to amend, while we shall indubitably ruin those of the Carthaginians, which have been hitherto so flourishing."

We are reminded by this of the story of Pericles, who, when ready to set sail with his fleet on a great expedition, saw himself stopped by a similar phenomenon. He spread his mantle over the eyes of the pilot, whom fear had prevented acting, and asked him if that was any sign of misfortune, when the pilot answered in the negative. "What misfortune then do you suppose," said he, "is presaged by the body that hides the sun, which differs from this in nothing but being larger?"

With reference to these eclipses, when their locality and approximate date is known, astronomy comes to the assistance of history, and can supply the exact day, and even hour, of the occurrence. For the eclipses depend on the motions of the moon, and just as astronomers can calculate both the time and the path of a solar eclipse in the future, so they can for the past. If then the eclipses are calculated back to the epoch when the particular one is recorded, it can be easily ascertained which one it was that about that time passed over the spot at which it was observed, and as soon as the particular eclipse is fixed upon, it may be told at what hour it would be seen.

Thus the eclipse of Thales has been assigned by different authors to various dates, between the 1st of October, 583 B.C., and the 3rd of February, 626 B.C. The only eclipse of the sun that is suitable between those dates has been found by the

Astronomer-Royal to be that which would happen in Lydia on the 28th of May, 585 B.C., which must therefore be the date of the event.

So of the eclipse of Agathocles, M. Delaunay has fixed its date to the 15th August, 310 B.C.

In later days, when Christopher Columbus had to deal with the ignorant people of America, the same kind of story was repeated. He found himself reduced to famine by the inhabitants of the country, who kept him and his companions prisoners; and being aware of the approach of the eclipse, he menaced them with bringing upon them great misfortunes, and depriving them of the light of the moon, if they did not instantly bring him provisions. They cared little for his menaces at first; but as soon as they saw the moon disappear, they ran to him with abundance of victuals, and implored pardon of the conqueror. This was on the 1st of March, 1504, a date which may be tested by the modern tables of the moon, and Columbus's account proved to be correct. The eclipse was indeed recorded in other places by various observers.

Eclipses in their natural aspect have thus had considerable influence on the vulgar, who knew nothing of their cause. This of course was the state with all in the early ages, and it is interesting to trace the gradual progress from their being quite unexpected to their being predicted.

It is very probable, if not certain, that their recurrence in

PLATE XIII.—CHRISTOPHER COLUMBUS AND THE ECLIPSE OF THE MOON.

the case of the moon at least was recognised long before their nature was understood.

Among the Chinese they were long calculated, and, in fact, it is thought by some that they have pretended to a greater antiquity by calculating backwards, and recording as observed eclipses those which happened before they understood or noticed them. It seems, however, authenticated that they did in the year 2169 B.C. observe an eclipse of the sun, and that at that date they were in the habit of predicting them. For this particular eclipse is said to have cost several of the astronomers their lives, as they had not calculated it rightly. As the lives of princes were supposed to be dependent on these eclipses, it became high treason to expose them to such a danger without forewarning them. They paid more attention to the eclipses of the sun than of the moon.

Among the Babylonians the eclipses of the moon were observed from a very early date, and numerous records of them are contained in the Observations of Bel in Sargon's library, the tablets of which have lately been discovered. In the older portion they only record that on the 14th day of such and such a (lunar) month an eclipse takes place, and state in what watch it begins, and when it ends. In a later portion the observations were more precise, and the descriptions of the eclipse more accurate. Long before 1700 B.C. the discovery of the lunar cycle of 223 lunar months had been made, and by means of it they were able to state of each lunar eclipse, that it was either "according to calculation" or "contrary to calculation."

They dealt also with solar eclipses, and tried to trace on a sphere the path they would take on the earth. Accordingly, like the eclipses of the moon, these too were spoken of as happening either "according to calculation" or "contrary to calculation." "In a report sent in to one of the later kings of Assyria by the state astronomer, Abil Islar states that a watch had been kept on the 28th, 29th, and 30th of Sivan, or May, for an eclipse of the sun, which did not, however, take place after all. The shadow, it is clear, must have fallen outside the field of observation." Besides the more ordinary kind of solar eclipses, mention is made in the Observations of Bel of annular eclipses which, strangely enough, are seldom alluded to by classical writers.

A record of a later eclipse has been found by Sir Henry Rawlinson on one of the Nineveh Tablets. This occurred near that city in B.C. 763, and from the character of the inscription it may be inferred that it was a rare occurrence with them, indeed that it was nearly, if not quite, a total eclipse. This has an especial interest as being the earliest that we have any approximate date for.

It is possible that the remarkable phenomenon, alluded to by the prophet Isaiah, of the shadow going backwards ten degrees on the dial of Ahaz, may be really a record of an eclipse of the sun, such as astronomy proves to have occurred at Jerusalem in the year 689 B.C.

We have very little notice of the calculation of eclipses by the Egyptians; all that is told us is more or less fabulous. Thus Diogenes Laertius says that they reckoned that during a period of 48,863 years, 373 eclipses of the sun and 832 eclipses of the moon had occurred, which is far fewer than the right number for so long a time, and which, of course, has no basis in fact.

Among the Greeks, Anaxagoras was the first who entertained clear ideas about the nature of eclipses; and it was from him that Pericles learnt their harmlessness.

Plutarch relates that Helicon of Cyzicus predicted an eclipse of the sun to Dionysius of Syracuse, and received as a reward a talent of silver.

Livy records an eclipse of the sun as having taken place on the 11th of Quintilis, which corresponds to the 11th of July. It happened during the Appollinarian games, 190 B.C.

The same author tells us of an eclipse of the moon that was predicted by one Gallus, a tribune of the second legion, on the eve of the battle of Pydna—a prediction which was duly fulfilled on the following night. The fact of its having been foretold quieted the superstitious fears of the soldiers, and gave them a very high opinion of Gallus. Other authors, among them Cicero, do not give so flattering a

story, but state that Gallus's part consisted only in explaining
the cause of the eclipse after it had happened. The date
of this eclipse was the 3rd of September, 168 B.C.

Ennius, writing towards the end of the second century
B.C., describes an eclipse which was said to have happened
nearly two hundred years before (404, B.C.), in the following
remarkable words :—" On the nones of July the moon
passed over the sun, and there was night." Aristarchus,
three centuries before Christ, understood and explained the
nature of eclipses ; but the chief of the ancient authors
upon this subject was Hipparchus. He and his disciples
were able to predict eclipses with considerable accuracy,
both as to their time and duration. Geminus and
Cleomedes were two other writers, somewhat later, who
explained and predicted eclipses. In later times regular
tables were drawn up, showing when the eclipses would
happen. One that Ptolemy was the author of was founded
on data derived from ancient observers—Callipus, Democritus, Eudoxus, Hipparchus—aided by his own calculations.
After the days of Ptolemy the knowledge of the eclipses
advanced *pari passu* with the advance of astronomy generally. So long as astronomy itself was empirical, the time of
the return of an eclipse was only reckoned by the intervals
that had elapsed during the same portion of previous cycles ;
but after the discovery of elliptic orbits and the force
of gravitation the whole motion of the moon could be

calculated with as great accuracy as any other astronomical phenomenon.

In point of fact, if the new moon is in the plane of the ecliptic there must be an eclipse of the sun; if the full moon is there, there must be an eclipse of the moon; and if it should in these cases be only partially in that plane, the eclipses also will be partial. The cycle of changes that the position of the moon can undergo when new and full occupies a period of eighteen years and eleven days, in which period there are forty-one eclipses of the sun and twenty-nine of the moon. Each year there are at most seven and at least two eclipses; if only two, they are eclipses of the sun. Although more numerous in reality for the whole earth, eclipses of the sun are more rarely observed in any particular place, because they are not seen everywhere, but only where the shadow of the moon passes; while all that part of the earth that sees the moon at all at the time sees it eclipsed.

We now come to comets.

The ancients divided comets into different classes, the chief points of distinction being derived from the shape, length, and brilliancy of the tails. Pliny distinguished twelve kinds, which he thus characterised :—" Some frighten us by their blood-coloured mane; their bristling hair rises towards the heaven. The bearded ones let their long hair fall down like a majestic beard. The javelin-shaped ones

seem to be projected forwards like a dart, as they rapidly
attain their shape after their first appearance; if the tail is
shorter, and terminates in a point, it is called a sword; this is
the palest of all the comets; it has the appearance of a bright
sword without any diverging rays. The plate or disc derives
its name from its shape, its colour is that of amber, it gives
out some diverging rays from its sides, but not in large
quantity. The cask has really the form of a cask, which
one might suppose to be staved in smoke enveloped in
light. The retort imitates the figure of a horn, and the
lamp that of a burning flame. The horse-comet represents
the mane of a horse which is violently agitated, as by a
circular, or rather cylindrical, motion. Such a comet appears
also of singular whiteness, with hair of a silver hue; it is
so bright that one can scarcely look at it. There are bristling
comets, they are like the skins of beasts with their hair on,
and are surrounded by a nebulosity. Lastly, the hair of the
comet sometimes takes the form of a lance."

Pingré, a celebrated historian of comets, tells us that one
of the first comets noticed in history is that which appeared
over Rome forty years before Christ, and in which the Roman
people imagined they saw the soul of Cæsar endowed with
divine honours. Next comes that which threw its light on
Jerusalem when it was being besieged and remained for a
whole year above the city, according to the account of
Josephus. It was of this kind that Pliny said it "is of so

great a whiteness that one can scarcely look at it, and one *may see in it the image of God in human form.*"

Diodorus tells us that, a little after the subversion of the towns of Helix and Bura, there were seen, for several nights in succession, a brilliant light, which was called a beam of fire, but which Aristotle says was a true comet.

Plutarch, in his life of Timoleon, says a burning flame preceded the fleet of this general until his arrival at Sicily, and that during the consulate of Caius Servilius a bright shield was seen suspended in the heavens.

The historians Sazoncenas and Socrates relate that in the year 400 A.D. a comet in the form of a sword shone over Constantinople, and appeared to touch the town just at the time when great misfortunes were impending through the treachery of Gainas.

The same phenomenon appeared over Rome previous to the arrival of Alaric.

In fact the ancient chroniclers always associated the appearance of a comet with some terrestrial event, which it was not difficult to do, seeing that critical situations were at all times existing in some one country or other where the comet would be visible, and probably those which could not be connected with any were not thought worthy of being recorded.

It is well known that the year 1000 A.D. was for a long time predicted to be the end of the world. In this year the

astronomers and chroniclers registered the fall of an enormous
burning meteor and the appearance of a comet. Pingré says :
" On the 19th of the calends of January "—that is the 14th of
December—" the heavens being dark, a kind of burning sword
fell to the earth, leaving behind it a long train of light. Its
brilliancy was such that it frightened not only those who
were in the fields, but even those who were shut up in their
houses. This great opening in the heavens was gradually
closed, and then was seen the figure of a dragon, whose feet
were blue, and whose head kept continually increasing. A
comet having appeared at the same time as this chasm, or
meteor, they were confounded." This relation is given in the
chronicles of Seigbert in Hermann Corner, in the Chronique
de Tours, in Albert Casin, and other historians of the time.

Bodin, resuscitating an idea of Democritus, wrote that the
comets were the souls of illustrious personages, who, after
having lived on the earth a long series of centuries, and being
ready at last to pass away, were carried in a kind of triumph
to heaven. For this reason, famine, epidemics, and civil wars
followed on the apparition of comets, the towns and their
inhabitants finding themselves then deprived of the help of
the illustrious souls who had laboured to appease their
intestinal feuds.

One of the comets of the middle ages which made the
greatest impression on the minds of the people was that
which appeared during Holy Week of the year 837, and

frightened Louis the Debonnaire. The first morning of its
appearance he sent for his astrologer. "Go," he said, "on to
the terrace of the palace, and come back again immediately
and tell me what you have seen, for I have not seen that star
before, and you have not shown it to me; but I know that this
sign is a comet: it announces a change of reign and the death
of a prince." The son of Charlemagne having taken counsel
with his bench of bishops, was convinced that the comet was
a notice sent from heaven expressly for him. He passed the
nights in prayer, and gave large donations to the monasteries,
and finally had a number of masses performed out of fear
for himself and forethought for the Church committed to his
care. The comet, however, was a very inoffensive one, being
none other than that known as Halley's comet, which returned in 1835. While they were being thus frightened in
France, the Chinese were observing it astronomically.

The historian of Merlin the enchanter relates that a few days
after the *fêtes* which were held on the occasion of the erection
of the funeral monument of Salisbury, a sign appeared in
heaven. It was a comet of large size and excessive splendour.
It resembled a dragon, out of whose mouth came a long two-
forked tongue, one part of which turned towards the north
and the other to the east. The people were in a state of
fear, each one asking what this sign presaged. Uter, in
the absence of the king, Ambrosius, his brother, who was
engaged in pursuing one of the sons of Vortigern, consulted

all the wise men of Britain, but no one could give him any
answer. Then he thought of Merlin the enchanter, and sent
for him to the court.   " What does this apparition presage ? "
demanded the king's brother. Merlin began to weep. "O son
of Britain, you have just had a great loss—the king is dead."
After a moment of silence he added, " But the Britons have
still a king. Haste thee, Uter, attack the enemy. All the
island will submit to you, for the figure of the fiery dragon
is thyself. The ray that goes towards Gaul represents a son
who shall be born to thee, who will be great by his achieve-
ments, and not less so by his power. The ray that goes
towards Ireland represents a daughter of whom thou shalt
be the father, and her sons and grandsons shall reign over all
the Britons."   These predictions were realised; but it is
more than probable that they were made up after the event.

The comet of 1066 was regarded as a presage of the
Conquest under William of Normandy. In the Bayeaux
tapestry, on which Matilda of Flanders had drawn all the
most memorable episodes in the transmarine expedition of
her husband, the comet appears in one of the corners with
the inscription, *Isti mirantur stellam*, which proves that the
comet was considered a veritable marvel. It is said even to
be traditionally reported that one of the jewels of the British
crown was taken from the tail of this comet. Nevertheless
it was no more than Halley's comet again in its periodical
visit every seventy-six years.

In July, 1214, a brilliant comet appeared which was lost to view on the same day as the Pope, Urban IV., died, *i.e.* the third of October.

In June, 1456, a similar body of enormous size, with a very long and extraordinarily bright tail, put all Christendom in a fright. The Pope, Calixtus III., was engaged in a war at that time with the Saracens. He showed the Christians that the comet "had the form of a cross," and announced some great event. At the same time Mahomet announced to his followers that the comet, "having the form of a yataghan," was a blessing of the Prophet's. It is said that the Pope afterwards recognised that it had this form, and excommunicated it. Nevertheless, the Christians obtained the victory under the walls of Belgrade. This was another appearance of Halley's comet.

In the early months of 1472 appeared a large comet, which historians agree in saying was very horrible and alarming. Belleforest said it was a hideous and frightening comet, which threw its rays from east to west, giving great cause for fear to great people, who were not ignorant that comets are the menacing rods of God, which admonish those who are in authority, that they may be converted.

Pingré, who has told us of so many of the comets that were seen before his time, wrote of this epoch: "Comets became the most efficacious signs of the most important and doubtful events. They were charged to announce wars,

seditions, and the internal movements of republics; they
presaged famines, pestilence, and epidemics; princes, or even
persons of dignity, could not pay the tribute of nature
without the previous appearance of that universal oracle,
a comet; men could no longer be surprised by any unexpected event; the future might be as easily read in the
heavens as the past in history. Their effect depended on the
place in the heavens where they shone, the countries over
which they directly lay, the signs of the zodiac that they
measured by their longitude, the constellations they traversed,
the form and length of their tails, the place where they went
out, and a thousand other circumstances more easily indicated
than distinguished; they also announced in general wars, and
the death of princes, or some grand personage, but there
were few years that passed without something of this kind
occurring. The devout astrologers—for there were many of
that sort—risked less than the others. According to them,
the comet threatened some misfortune; if it did not happen,
it was because the prayers of penitence had turned aside the
wrath of God; he had returned his sword to the scabbard.
But a rule was invented which gave the astrologers free
scope, for they said that events announced by a comet might
be postponed for one or more periods of forty years, or even
as many years as the comet had appeared days; so that one
which had appeared for six months need not produce its
effect for 180 years."

The most frightful of the comets of this period, according
to Simon Goulart, was that of 1527. "It put some into so
great a fright that they died; others fell sick. It was seen
by several thousand people, and appeared very long, and of
the colour of blood. At the summit was seen the representa-

Fig. 55.—Representation of a Comet, 16th Century.

tion of a curved arm, holding a large sword in its hand, as
if it would strike; at the top of the point of the sword
were three stars, but that which touched the point was
more brilliant than the others. On the two sides of the
rays of this comet were seen large hatchets, poignards,

bloody swords, among which were seen a great number of men decapitated, having their heads and beards horribly bristling."

A view of this comet is given in the *History of Prodigies*.

There was another comet remarked in 1556, and another in 1577, like the head of an owl, followed by a mantle of scattered light, with pointed ends. Of this comet we read in the same book that recorded the last described: "The comet is an infallible sign of a very evil event. Whenever eclipses of the sun or moon, or comets, or earthquakes, conversions of water into blood, and such like prodigies happen, it has always been known that very soon after these miserable portents afflictions, effusion of human blood, massacres, deaths of great monarchs, kings, princes, and rulers, seditions, treacheries, raids, overthrowings of empires, kingdoms, or villages; hunger and scarcity of provisions, burning and overthrowing of towns; pestilences, widespread mortality, both of beasts and men; in fact all sorts of evils and misfortunes take place. Nor can it be doubted that all these signs and prodigies give warning that the end of the world is come, and with it the terrible last judgment of God."

But even now comets were being observed astronomically, and began to lose their sepulchral aspect.

A remarkable comet, however, which appeared in 1680, was not without its fears for the vulgar. We are told that it was recognised as the same which appeared the year of

Cæsar's death, then in 531, and afterwards in 1106, having a period of about 575 years. The terror it produced in the towns was great; timid spirits saw in it the sign of a new deluge, as they said water was always announced by fire. While the fearful were making their wills, and, in anticipation of the end of the world, were leaving their money to the monks, who in accepting them showed themselves better physicists than the testators, people in high station were asking what great person it heralded the death of, and it is reported of the brother of Louis XIV., who apparently was afraid of becoming too suddenly like Cæsar, that he said sharply to the courtiers who were discussing it, " Ah, gentlemen, you may talk at your ease, if you please; you are not princes."

This same comet gave rise to a curious story of an " extraordinary prodigy, how at Rome a hen laid an egg on which was drawn a picture of the comet.

' The fact was attested by his Holiness, by the Queen of Sweden, and all the persons of first quality in Rome. On the 4th December, 1680, a hen laid an egg on which was seen the figure of the comet, accompanied by other marks such as are here represented. The cleverest naturalists in Rome have seen and examined it, and have never seen such a prodigy before."

Of this same comet Bernouilli wrote, "*That if the body of the comet is not a visible sign of the anger of God, the tail may*

le."   It was this too that suggested to Whiston the idea that
he put forward, not as a superstitious, but as a physical
speculation, that a comet approaching the earth was the cause
of the deluge.

FIG. 56. —AN EGG MARKED WITH A COMET.

The last blow to the superstitious fear of the comets was
given by Halley, when he proved that they circulated like
planets round the sun, and that the comets noticed in 837,
1066, 1378, 1456, 1531, 1607, 1682 were all one, whose period
was about 76 years, and which would return in 1759, which

prediction was verified, and the comet went afterwards by the name of this astronomer. It returned again in 1835, and will revisit us in 1911.

Even after the fear arising from the relics of astrology had died away, another totally different alarm was connected with comets—an alarm which has not entirely subsided even in our own times. This is that a comet may come in contact with the earth and destroy it by the collision. The most remarkable panic in this respect was that which arose in Paris in 1773. At the previous meeting of the Academy of Sciences, M. Lalande was to have read an interesting paper, but the time failed. It was on the subject of comets that could, by approaching the earth, cause its destruction, with special reference to the one that was soon to come. From the title only of the paper the most dreadful fears were spread abroad, and, increasing day by day, were with great difficulty allayed. The house of M. Lalande was filled with those who came to question him on the memoir in question. The fermentation was so great that some devout people, as ignorant as weak, asked the archbishop to make a forty hours' prayer to turn away the enormous deluge that they feared, and the prelate was nearly going to order these prayers, if the members of the Academy had not persuaded him how ridiculous it would be. Finally, M. Lalande, finding it impossible to answer all the questions put to him about his fatal memoir, and wishing to prevent the real evils that

A A

might arise from the frightened imaginations of the weak, caused it to be printed, and made it as clear as was possible. When it appeared, it was found that he stated that of the sixty comets known there were eight which could, by coming too near the earth, say within 40,000 miles, occasion such a pressure that the sea would leave its bed and cover part of the globe, but that in any case this could not happen till after twenty years. This was too long to make it worth while to make provision for it, and the effervescence subsided.

A similar case to this occurred with respect to Biela's comet, which was to return in 1832. In calculating its re-appearance in this year, Damoiseau found that it would pass through the plane of the earth's orbit on the 29th of October. Rushing away with this, the papers made out that a collision was inevitable, and the end of the world was come. But no one thought to inquire where the earth would be when the comet passed through the plane in which it revolved. Arago, however, set people's minds at rest by pointing out that at that time the earth would be a month's journey from the spot, which with the rate at which the earth is moving would correspond to a distance of sixty millions of miles.

This, like other frights, passed away, but was repeated again in 1840 and 1857 with like results, and even in 1872 a similar end to the world was announced to the public for

the 12th of August, on the supposed authority of a Professor
at Geneva, but who had never said what was supposed.

But in reality all cause of fear has now passed away,
since it has been proved that the comet is made of gaseous
matter in a state of extreme tenuity, so that, though it may
make great show in the heavens, the whole mass may not
weigh more than a few pounds; and we have in addition
the testimony of experience, which might have been relied
on on the occasions above referred to, for in 1770 Lexele's
comet was seen to pass through the satellites of Jupiter
without deranging them in the least, but was itself thrown
entirely out of its path, while there is reason to believe that
on the 29th of June, 1861, the earth remained several hours
in the tail of a comet without having experienced the
slightest inconvenience.

As to the nature of comets, the opinions that have been
held have been mostly very vague. Metrodorus thought
they were reflections from the sun; Democritus, a concourse
of several stars; Aristotle, a collection of exhalations which
had become dry and inflamed; Strabo, that they are the
splendour of a star enveloped in a cloud; Heracletes of
Pontus, an elevated cloud which gave out much light;
Epigenes, some terrestrial matter that had caught fire, and
was agitated by the wind; Bœcius, part of the air, coloured;
Anaxagoras, sparks fallen from the elementary fire; Xenophanes, a motion and spreading out of clouds which caught

fire; and Descartes, the débris of vortices that had
been destroyed, the fragments of which were coming
towards us.

It is said that the Chaldæans held the opinion that they
were analogous to planets by their regular course, and that
when we ceased to see them, it was because they had
gone too far from us; and Seneca followed this explanation,
since he regarded them as globes turning in the heavens,
and which appear and disappear in certain times, and
whose periodical motions might be known by regular
observation.

We have thus traced the particular ideas that have
attached themselves to eclipses and comets, as the two
most remarkable of the extraordinary phenomena of the
heavens, and have seen how the fears and superstitions of
mankind have been inevitably linked with them in the
earlier days of ignorance and darkness, but they are only
part of a system of phenomena, and have been no more
connected with superstition than others less remarkable,
except in proportion to their remarkableness. Other minor
appearances that are at all unusual have, on the same
belief in the inextricable union of celestial and terrestrial
matters, been made the signs of calamities or extra-pros-
perity; the doleful side of human nature being usually the
strongest, the former have been chosen more often than the
latter.

According to Seneca, the tradition of the Chaldees announced that a universal deluge would be caused by the conjunction of all the planets in the sign of Capricorn, and that a general breaking up of the earth would take place at the moment of their conjunction in Cancer. " The general break-up of the world," they said, " will happen when the stars which govern the heaven, penetrated with a quality of heat and dryness, meet one another in a fiery triplicity."

Everywhere, and in all ages of the past, men have thought that a protecting providence, always watching over them, has taken care to warn them of the destinies which await them ; thence the good and evil *presages* taken from the appearance of certain heavenly bodies, of divers meteors, or even the accidental meeting of certain animate or in-animate objects. The Indian of North America dying of famine in his miserable cabin, will not go out to the chase if he sees certain presages in the atmosphere. Nor need we be astonished at such ideas in an uncultivated man, when even among Europeans, a salt-cellar upset, a glass broken, a knife and fork crossed, the number thirteen at dinner, and such things are regarded as unlucky accidents. The employment of sorcery and divination is closely connected with these superstitions. Besides eclipses and comets, meteors were taken as the signs of divine wrath. We learn from S. Maximus of Turin, that the Christians of his time admitted the necessity of making a noise during eclipses,

PLATE XIV.—PRODIGIES IN THE MIDDLE AGES.

so as to prevent the magicians from hurting the sun or moon, a superstition entirely pagan. They used to fancy they could see celestial armies in the air, coming to bring miraculous assistance to man. They thought the hurricanes and tempests the work of evil spirits, whose rage kept them set against the earth. S. Thomas Aquinas, the great theologian of the thirteenth century, accepted this opinion, just as he admitted the reality of sorceries. But the full development, as well as the nourishment of these superstitious ideas, was derived from the storehouse of astrology, which dealt with matters of ordinary occurrence, both in the heavens and on the earth—and to the history of which our next chapter is devoted.

CHAPTER XIII

OUR study of the opinions of the ancients on the various phenomena of astronomy, leads us inevitably to the discussion of their astrology, which has in every age and among every people accompanied it—and though astrology be now no more as a science, or lingers only with those who are ignorant and desirous of taking advantage of the still greater ignorance of others—yet it is not lacking in interest as showing the effect of the phenomena of the heavens on the human mind, when that effect is brought to its most technical and complete development.

We must distinguish in the first place two kinds of astrology, viz., natural and judicial. The first proposed to foresee and announce the changes of the seasons, the rains, wind, heat, cold, abundance, or sterility of the ground, diseases, &c., by means of a knowledge of the causes which act on the air and on the atmosphere. The other is occupied with objects which would be still more interesting to men. It traced at

the moment of his birth, or at any other period of his life,
the line that each must travel according to his destiny.   It
pretended to determine our characters, our passions, fortune,
misfortunes, and perils in reserve for each mortal.

We have not here to consider the natural astrology, which
is a veritable science of observation and does not deserve the
name of astrology.   It is rather worthy to be called the
meteorological calendar of its cultivators.   More rural than
their descendants of the nineteenth century, the ancients
had recognised the connection between the celestial phenomena and the vicissitudes of the seasons; they observed
these phenomena carefully to discover the return of the
same inclemencies; and they were able (or thought they
were) to state the date of the return of particular kinds of
weather with the same positions of the stars.   But the very
connection with the stars soon led the way to a degeneracy.
The autumnal constellations, for example, Orion and Hercules,
were regarded as rainy, because the rains came at the time
when these stars rose.   The Egyptians who observed in the
morning, called Sirius "the burning," because his appearance
in the morning was followed by the great heat of the summer:
and it was the same with the other stars.   Soon they regarded
them as the cause of the rain and the heat—although they
were but remote witnesses.   The star Sirius is still connected
with heat—since we call it the dog-star—and the hottest days
of the year, July 22nd to August 23rd, we call dog-days.

At the commencement of our era, the morning rising of Sirius took place on the earlier of those days—though it does not now rise in the morning till the middle of August—and 4,000 years ago it rose about the 20th of June, and preceded the annual rise of the Nile.

The belief in the meteorological influence of the stars is one of the causes of judicial astrology. This latter has simply subjected man, like the atmosphere, to the influence of the stars; it has made dependent on them the risings of his passions, the good and ill fortune of his life, as well as the variations of the seasons. Indeed, it was very easy to explain. It is the stars, or heavenly bodies in general, that bring the winds, the rains, and the storms; their influences mixed with the action of the rays of the sun modify the cold or heat; the fertility of the fields, health or sickness, depend on these beneficial or injurious influences; not a blade of grass can grow without all the stars having contributed to its increase; man breathes the emanations which escaping from the heavenly bodies fill the air; man is therefore in his entire nature subjected to them; these stars must therefore influence his will and his passions; the good and evil passages in his career, in a word, must direct his life.

As soon as it was established that the rising of a certain star or planet, and its aspect with regard to other planets, announced a certain destiny to man, it was natural to believe that the rarer configurations signified extraordinary events, which concerned great empires, nations, and towns. And lastly, since errors grow faster than truth, it was natural to think that the configurations which were still more rare, such as the reunion of all the planets in conjunction with the same star, which can occur only after thousands of centuries, while nations have been renewed an infinity of times, and empires have been ruined, had reference to the earth itself, which had served as the theatre for all these events. Joined to these superstitious ideas was the tradition of a deluge, and the belief that the world must one day perish by fire, and so it was announced that the former event took place when all the planets were in conjunction in the sign of the Fishes, and the latter would occur when they all met in the sign of the Lion.

The origin of astrology, like that of the celestial sphere, was in all probability in upper Asia.

There, the starry heavens, always pure and splendid, invited observation and struck the imagination. We have already seen this with respect to the more matter-of-fact portions of astronomy. The Assyrians looked upon the stars as divinities endued with beneficent or maleficent power. The adoration of the heavenly bodies was the earliest form of religion among the pastoral population that came down from the mountains of Kurdestan to the plains of Babylon. The Chaldæans at last set apart a sacerdotal and learned caste

devoted to the observation of the heavens; and the temples
became regular observatories. Such doubtless was the tower
of Babel—a monument consecrated to the seven planets, and
of which the account has come down to us in the ancient
book of Genesis.

A long series of observations put the Chaldæans in possession of a theological astronomy, resting on a more or less
chimerical theory of the influence of the celestial bodies on
the events of nations and private individuals. Diodorus
Siculus, writing towards the commencement of our era, has
put us in possession of the most circumstantial details that
have reached us with regard to the Chaldæan priests.

At the head of the gods, the Assyrians placed the sun and
moon, whose courses and daily positions they had noted in
the constellation of the zodiac, in which the sun remained,
one month in each. The twelve signs were governed by as
many gods, who had the corresponding months under their
influence. Each of these months were divided into three
parts, which made altogether thirty-six subdivisions, over
which as many stars presided, called gods of consultation.
Half of these gods had under their control the things which
happen above the earth, and the other half those below. The
sun and moon and the five planets occupied the most elevated
rank in the divine hierarchy and bore the name of gods of
interpretation. Among these planets Saturn or old Bel,
which was regarded as the highest star and the most distant

from us, was surrounded by the greatest veneration; he was the interpreter *par excellence*—the revealer. Each of the other planets had his own particular name. Some of them, such as *Bel* (Jupiter), *Merodaez* (Mars), *Nebo* (Mercury), were regarded as male, and the others, as *Sin* (the Moon), and *Mylitta* or *Baulthis* (Venus), as females; and from their position relative to the zodiacal constellations, which were also called *Lords* or masters of the *Gods*, the Chaldæans derived the knowledge of the destiny of the men who were born under such and such a conjunction—predictions which the Greeks afterwards called horoscopes. The Chaldæans invented also relations between each of the planets and meteorological phenomena, an opinion partly founded on fortuitous coincidences which they had more or less frequently observed. In the time of Alexander their credit was considerable, and the king of Macedonia, either from superstition or policy, was in the habit of consulting them.

It is probable that the Babylonian priests, who referred every natural property to sidereal influences, imagined there were some mysterious relations between the planets and the metals whose colours were respectively somewhat analogous to theirs. Gold corresponded to the sun, silver to the moon, lead to Saturn, iron to Mars, tin to Jupiter, and mercury still retains the name of the planet with which it was associated. It is less than two centuries ago, since the metals have ceased to be designated by the signs of their respective planets. Alchemy, the mother of chemistry, was an intimately connected sister of Astrology, the mother of Astronomy.

Egyptian civilisation dates back to a no less remote period than that of Babylon. Not less careful observers than the Babylonish astrologers of the meteors and the atmospheric revolutions, they could predict certain phenomena, and they gave it out that they had themselves been the cause of them.

Diodorus Siculus tells us that the Egyptian priests pretty generally predicted the years of barrenness or abundance, the contagions, the earthquakes, inundations, and comets. The knowledge of celestial phenomena made an essential part of the theology of the Egyptians as it did of the Chaldæans. They had colleges of priests specially attached to the study of the stars, at which Pythagoras, Plato, and Eudoxus were instructed.

Religion was besides completely filled with the symbols relating to the sun or moon. Each month, each decade, each day was consecrated to a particular god. These gods, to the number of thirty, were called in the Alexandrine astronomy *decans* (δέκανοι). The festivals were marked by the periodical return of certain astronomical phenomena, and those heliacal risings to which any mythological ideas were attached, were noted with great care. We find even now proof of this old sacerdotal science in the zodiac sculptured on the ceilings of certain temples, and in the hieroglyphic inscriptions relating to celestial phenomena.

According to the Egyptians, who were no less aware than the Greeks, of the influence of atmospheric changes on our organs, the different stars had a special action on each part of the body. In the funeral rituals which were placed at the bottom of the coffins, constant allusion is made to this theory. Each limb of the dead body was placed under the protection of a particular god. The divinities divided between them, so to speak, the spoils of the dead. The head belonged to Ra, or the Sun, the nose and lips to Anubis, and so on. To establish the horoscope of anyone, this theory of specific influences was combined with the state of the heavens at the time of his birth. It seems even to have been the doctrine of the Egyptians, that a particular star indicated the coming of each man into the world, and this opinion was held also by the Medes, and is alluded to in the Gospels. In Egypt, as in Persia and Chaldæa, the science of nature was a sacred doctrine, of which magic and astrology constituted the two branches, and in which the phenomena of the universe were attached very firmly to the divinities or genii with which they believed it filled. It was the same in the primitive religions of Greece.

The Thessalian women had an especially great reputation in the art of enchantments. All the poets rival one another in declaring how they are able, by their magical hymns, to bring down the moon. Menander, in his comedy entitled *The Thessalian*, represents the mysterious ceremonies by the aid of which these sorcerers force the moon to leave the heavens, a prodigy which so completely became the type of enchantments that Nonnus tells us it is done by the Brahmins. There was, in addition, another *cultus* in Greece, namely, that of Hecate with mysterious rays, the patron of sorcerers. Lucian of Samosate—if the work on astrology which is ascribed to him be really his—justifies his belief in the influence of the stars in the following terms :—" The stars follow their orbit in the heaven ; but independently of their motion, they act upon what passes here below. If you admit that a horse in a gallop, that birds in flying, and men in walking, make the stones jump or drive the little floating particles of dust by the wind of their course, why should you deny that the stars have any effect ? The smallest fire sends us its emanations, and although it is not for us that the stars burn, and they care very little about warming us, why should we not receive any emanations from them? Astrology, it is true, cannot make that good which is evil. It can effect no change in the course of events, but it renders a service to those who cultivate it by announcing to them good things to come ; it procures joy by anticipation at the same time that it fortifies them against the evil. Misfortune, in fact, does not take them by surprise, the foreknowledge of it renders it easier and lighter. That is my way of looking at astrology."

Very different is the opinion of the satirist Juvenal, who says that women are the chief cultivators of it. "All that an astrologer predicts to them," he says, "they think to come from the temple of Jupiter. Avoid meeting with a lady who is always casting up her *ephemerides*, who is so good an astrologer that she has ceased to consult, and is already beginning to be consulted; such a one on the inspection of the stars will refuse to accompany her husband to the army or to his native land. If she only wishes to drive a mile, the hour of departure is taken from her book of astrology. If her eye itches and wants rubbing, she will do nothing till she has run through her conjuring book. If she is ill in bed, she will take her food only at the times fixed in her *Petosiris*. Women of second-rate condition," he adds, "go round the circus before consulting their destiny, after which they show their hands and face to the diviner."

When Octavius came into the world a senator versed in astrology, Nigidius Figulus, predicted the glorious destiny of the future emperor. Livia, the wife of Tiberius, asked another astrologer, Scribius, what would be the destiny of her infant; his reply was, they say, like the other's.

The house of Poppea, the wife of Nero, was always full of astrologers. It was one of the soothsayers attached to her house, Ptolemy, who predicted to Otho his elevation to the empire, at the time of the expedition into Spain, where he accompanied him.

The history of astrology under the Roman empire supplies some very curious stories, of which we may select an illustrative few.

Octavius, in company with Agrippa, consulted one day the astrologer Theagenes. The future husband of Julia, more credulous or more curious than the nephew of Cæsar, was the first to take the horoscope. Theagenes foretold astonishing prosperity for him. Octavius, jealous of so happy a destiny, and fearing that the reply would be less favourable to him, instead of following the example of his companion, refused at first to state the day of his birth. But, curiosity getting the better of him, he decided to reply. No sooner had he told the day of his birth than the astrologer threw himself at his feet, and worshipped him as the future master of the empire. Octavius was transported with joy, and from that moment was a firm believer in astrology. To commemorate the happy influence of the zodiacal sign under which he was born, he had the picture of it struck on some of the medals that were issued in his reign.

The masters of the empire believed in astrological divination, but wished to keep the advantages to themselves. They wanted to know the future without allowing their subjects to do the same. Nero would not permit anyone to study philosophy, saying it was a vain and frivolous thing, from which one might take a pretext to divine future events. He feared lest some one should push his curiosity so far as to wish to find out when and how the emperor should die—a sort of indiscreet question, replies to which lead to conspiracies and attempts. This was what the heads of the state were most afraid of.

Tiberius had been to Rhodes, to a soothsayer of renown, to instruct himself in the rules of astrology. He had attached to his person the celebrated astrologer Thrasyllus, whose fate-revealing science he proved by one of those pleasantries which are only possible with tyrants.

Whenever Tiberius consulted an astrologer he placed him in the highest part of his palace, and employed for his purpose an ignorant and powerful freedman, who brought by difficult paths, bounded by precipices, the astrologer whose science his Majesty wished to prove. On the return journey, if the astrologer was suspected of indiscretion or treachery, the freedman threw him into the sea, to bury the secret. Thrasyllus having been brought by the same route across these precipices, struck Tiberius with awe while he questioned him, by showing him his sovereign power, and easily disclosing the things of the future Cæsar asked him if he had taken his own horoscope, and with what signs were marked that day and hour for himself. Thrasyllus then examined the position and the distance of the stars; he hesitated at first, then he grew pale; then he looked again, and finally, trembling with astonishment and fear, he cried out that the moment was perilous, and he was very near his last hour. Tiberius then embraced him and congratulated him on having escaped a danger by foreseeing it; and accepting henceforth all his predictions as oracles, he admitted him to the number of his intimate friends.

Tiberius had a great number of people put to death who were accused of having taken their horoscope to know what honours were in store for them, although in secret he took the horoscopes of great people, that he might ascertain that he had no rivalry to fear from them. Septimus Severus was very nearly paying with his head for one of those superstitious curiosities that brought the ambitious of the time to the astrologer. In prosperous times he had gained faith in their predictions, and consulted them about important acts. Having lost his wife, and wishing to contract a second marriage, he took the horoscopes of the well-connected ladies who were at the time open to marriage. None of their fortunes, taken by the rules of astrology, were encouraging. He learnt at last that there was living in Syria a young woman to whom the Chaldæans had predicted that she should be the wife of a king. Severus was as yet but a legate. He hastened to demand her in marriage, and he obtained her; Julia was the name of the woman who was born under so happy a star; but was he the crowned husband which the stars had promised to the young Syrian? This reflection soon began to perplex

Severus, and to get out of his perplexity he went to Sicily
to consult an astrologer of renown. The matter came to
the ears of the Emperor Commodus ; and judge of his anger !
The anger of Commodus was rage and frenzy; but the
event soon gave the response that Severus was seeking
in Sicily,—Commodus was strangled.

Divination which had the emperor for its object at last
came to be a crime of high treason. The rigorous measures
resorted to against the indiscreet curiosity of ambition
took more terrible proportions under the Christian emperors.

Under Constantine, a number of persons who had applied
to the oracles were punished with cruel tortures.

Under Valens, a certain Palladius was the agent of a terrible persecution. Everyone found himself exposed to being
denounced for having relations with soothsayers. Traitors
slipped secretly into houses magic formulæ and charms,
which then became so many proofs against the inhabitant.
The fear was so great in the East, says Ammienus Marcellinus,
that a great number burned their books, lest matter should
be found in them for an accusation of magic or sorcery.

One day in anger, Vitellius commanded all the astrologers
to leave Italy by a certain day. They responded by a poster,
which impudently commanded the prince to leave the
earth before that date, and at the end of the year Vitellius
was put to death ; on the other hand, the confidence accorded
to astrologers led sometimes to the greatest extremes.

For instance, after having consulted Babylus, Nero put to death all those whose prophecies promised the elevation of Heliogabalus. Another instance was that of Marcus Aurelius and his wife Faustina. The latter was struck with the beauty of a gladiator. For a long time she vainly strove in secret with the passion that consumed her, but the passion did nothing but increase. At last Faustina revealed the matter to her husband, and asked him for some remedy that should restore peace to her troubled soul. The philosophy of Marcus Aurelius could not suggest anything. So he decided to consult the Chaldæans, who were adepts at the art of mixing philters and composing draughts. The means prescribed were more simple than might have been expected from their complicated science; it was that the gladiator should be cut in pieces. They added that Faustina should afterwards be anointed with the blood of the victim. The remedy was applied, the innocent athlete was immolated, and the empress afterwards only dreamed of him with great pleasure.

The first Christians were as much addicted to astrology as the other sects. The Councils of Laodicea (366, A.D.), of Arles (314), of Agdus (505), Orleans (511), Auxerre (570), and Narbonne (589), condemned the practice. According to a tradition of the commencement of our era, which appears to have been borrowed from Mazdeism, it was the rebel angels who taught men astrology and the use of charms.

Under Constantius the crime of high treason served as a pretext for persecution. A number of people were accused of it, who simply continued to practise the ancient religion. It was pretended that they had recourse to sorceries against the life of the emperor, in order to bring about his fall. Those who consulted the oracles were menaced with severe penalties and put to death by torture, under the pretence that by dealing with questions of fate they had criminal intentions. Plots without number multiplied the accusations; and the cruelty of the judges aggravated the punishments. The pagans in their turn had to suffer the martyrdom which they had previously inflicted on the early disciples of Christ—or rather, to be truer, it was authority, always intolerable, whether pagan or Christian, that showed itself inexorable against those who dared to differ from the accepted faith. Libanius and Jamblicus were accused of having attempted to discover the name of the successor to the empire. Jamblicus, being frightened at the prosecution brought against him, poisoned himself. The name only of philosopher was sufficient to found an accusation upon. The philosopher Maximus Diogenes Alypius, and his son Hierocles, were condemned to lose their lives on the most frivolous pretence. An old man was put to death because he was in the habit of driving off the approach of fever by incantations, and a young man who was surprised in the act of putting his hands alternately to a marble and his breast, because he thought that by counting in this way seven times seven, he might cure the stomach-ache, met with the same fate.

Theodosius prohibited every kind of manifestation or usage connected with pagan belief. Whoever should dare to immolate a victim, said his law, or consult the entrails of the animals he had killed, should be regarded as guilty of the crime of high treason.

The fact of having recourse to a process of divination was sufficient for an accusation against a man.

Theodosius II. thought that the continuation of idolatrous practices had drawn down the wrath of heaven, and brought upon them the recent calamities that had afflicted his empire—the derangement of the seasons and the sterility of the soil—and he thundered out terrible threats when his faith and his anger united themselves into fanaticism.

He wrote as follows to Florentius, prefect of the prætorium in 439, the year that preceded his death :—

"Are we to suffer any longer from the seasons being upset by the effect of the divine wrath, on account of the atrocious perfidy of the pagans, which disturbs the equilibrium of nature? For what is the cause that now the spring has no longer its ordinary beauty, that the autumn no longer furnishes a harvest to the laborious workman and that the winter, by its rigour, freezes the soil and renders it sterile?"

Perhaps we are unduly amused with these ideas of Theodosius so long as we retain the custom of asking the special in-
tervention of Providence for the presence or absence of rain !

In the middle ages, when astrology took such a hold on the
world, several philosophers went so far as to consider the
celestial vault as a book, in which each star, having the value
of one of the letters of the alphabet, told in ineffaceable
characters the destiny of every empire. The book of *Unheard-
of Curiosities*, by Gaffarel, gives us the configuration of these
celestial characters, and we find them also in the writings of
Cornelius Agrippa. The middle ages took their astrological
ideas from the Arabians and Jews. The Jews themselves at
this epoch borrowed their principles from such contaminated
sources that we are not able to trace in them the transmission
of the ancient ideas. To give an example, Simeon Ben-Jochai,
to whom is attributed the famous book called *Zohar*, had
attained in their opinion such a prodigious acquaintance with
celestial mysteries as indicated by the stars, that he could
have read the divine law in the heavens before it had been
promulgated on the earth. During the whole of the middle
ages, whenever they wanted to clear up doubts about geography or astronomy, they always had recourse to this
Oriental science, as cultivated by the Jews and Arabians.
In the thirteenth century Alphonse X. was very importunate
with the Jews to make them assist him with their advice in
his vast astronomical and historical works.

Nicholas Oresmus, when the most enlightened monarch in Europe was supplying Du Guesclin with an astrologer to guide him in his strategical operations, was physician to Charles V. of France, who was himself devoted to astrology, and gave him the bishopric of Lisieux. He composed the *Treatise of the Sphere*, of which we have already spoken. A few years later, a learned man, the bishop Peter d'Ailly, actually dared to take the horoscope of Jesus Christ, and proved by most certain rules that the great event which inaugurated the new era was marked with very notable signs in astrology.

Mathias Corvin, King of Hungary, never undertook anything without first consulting the astrologers. The Duke of Milan and Pope Paul also governed themselves by their advice. King Louis XI., who so heartily despised the rest of mankind, and had as much malice in him as he had weakness, had a curious adventure with an astrologer.

It was told him that an astrologer had had the hardihood to predict the death of a woman of whom the king was very fond. He sent for the wretched prophet, gave him a severe reprimand, and then asked him the question, "You, who know everything, when will *you* die?" The astrologer, suspecting a trick, replied immediately, "Sire, three days before your Majesty." Fear and superstition overcame the monarch's resentment, and the king took particular care of the adroit impostor.

It is well known how much Catherine de Medicis was under the influence of the astrologers. She had one in her Hôtel de Soissons in Paris, who watched constantly at the top of a tower. This tower is still in existence, by the Wool-Market, which was built in 1763 on the site of the hotel. It is surmounted by a sphere and a solar dial, placed there by the astronomer Pingré.

One of the most celebrated of the astrologers who was under her patronage was Nostradamus. He was a physician of Provence, and was born at St. Reny in 1503. To medicine he joined astrology, and undertook to predict future events. He was called to Paris by Catherine in 1556, and attempted to write his oracles in poetry. His little book was much sought after during the whole of the remainder of the sixteenth century, and even in the beginning of the next. According to contemporary writers many imitations were made of it. It was written in verses of four lines, and was called *Quatrains Astronomiques*. As usual, the prophecies were obscure enough to suit anything, and many believers have thought they could trace in the various verses prophecies of known events, by duly twisting and manipulating the sense.

A very amusing prophecy, which happened to be too clear to leave room for mistakes as to its meaning, and which turned out to be most ludicrously wrong, was one contained in a little book published in 1572 with this title:—*Prognos-tication touching the marriage of the very honourable and*

beloved Henry, by the Grace of God King of Navarre, and
the very illustrious Princess Marguerite of France, calculated
by Master Bernard Abbatio, Doctor in Medicine, and Astrologer
to the very Christian King of France.

First he asked if the marriage would be happy, and
says:—"Having in my library made the figure of the
heavens, I found that the lord of the ascendant is joined
to the lord of the seventh house, which is for the woman
of a trine aspect, from whence I have immediately concluded,
according to the opinion of Ptolemy, Haly, Zael, Messahala,
and many other sovereign astrologers, that they will love
one another intensely all their lives." In point of fact they
always detested each other. Again, "as to length of
life, I have prepared another figure, and have found that
Jupiter and Venus are joined to the sun with fortification,
and that they will approach a hundred years;" after all
Henri IV. died before he was sixty. "Our good King of
Navarre will have by his most noble and virtuous Queen
many children; since, after I had prepared another figure of
heaven, I found the ascendant and its lord, together with
the moon, all joined to the lord of the fifth house, called
that of children, which will be pretty numerous, on account
of Jupiter and also of Venus;" and yet they had no children!
"Jupiter and Venus are found domiciled on the aquatic
signs, and since these two planets are found concordant
with the lord of the ascendant, all this proves that the

children will be upright and good, and that they will love their father and mother, without doing them any injury, nor being the cause of their destruction, as is seen in the fruit of the nut, which breaks, opens, and destroys the stock from which it took its birth. The children will live long, they will be good Christians, and with their father will make themselves so benign and favourable towards those of our religion, that at last they will be as beloved as any man of our period, and there will be no more wars among the French, as there would have been but for the present marriage. God grant us grace that so long as we are in this transitory life we may see no other king but Charles IX., the present King of France." And yet these words were written in the year of the massacre of St. Bartholomew's day! and the marriage was broken off, and Henri IV. married to Marie de Medici. So much for the astrological predictions!

The aspect in which astrology was looked upon by the better minds even when it was flourishing may be illustrated by two quotations we may make, from Shakespeare and Voltaire.

Our immortal poet puts into the mouth of Edmund in *King Lear* :—"This is the excellent foppery of the world, that when we are sick in fortune (often the surfeit of our own behaviour) we make guilty of our disasters the sun, the moon, and the stars, as if we were villains by necessity; fools by heavenly compulsion; knaves, thieves, and treacherous, by spherical predominance; drunkards, liars, and adulterers by an enforced obedience of planetary influence; and all that we are evil in, by a divine thrusting on.   An admirable evasion of a libertine to lay his goatish disposition to the charge of a star!   My father married my mother under the Dragon's tail; and my nativity was under *Ursa major;* so that it follows I am rough lecherous.   Tut, I should have been that I am, had the maidenliest star in the firmament twinkled on my birth."

Voltaire writes thus :—"This error is ancient, and that is enough.   The Egyptians, the Chaldæans, the Jews could predict, and therefore we can predict now.   If no more predictions are made it is not the fault of the art.   So said the alchemists of the philosopher's stone.   If you do not find to-day it is because you are not clever enough ; but it is certain that it is in the clavicle of Solomon, and on that certainty more than two hundred families in Germany and France have been ruined.   Do you wonder either that so many men, otherwise much exalted above the vulgar, such as princes or popes, who knew their interests so well, should be so ridiculously seduced by this impertinence of astrology.   They were very proud and very ignorant.   There were no stars but for them ; the rest of the universe was *canaille,* for whom the stars did not trouble themselves.   I have not the honour of being a prince.   Nevertheless, the celebrated

Count of Boulainvilliers and an Italian, called Colonne, who had great reputation in Paris, both predicted to me that I should infallibly die at the age of thirty-two. I have had the malice already to deceive them by thirty years, for which I humbly beg their pardon."

The method by which these predictions were arrived at consisted in making the different stars and planets responsible for different parts of the body, different properties, and different events, and making up stories from the association of ideas thus obtained, which of course admitted of the greatest degree of latitude. The principles are explained by Manilius in his great poem entitled *The Astronomicals*, written two thousand years ago.

According to him the sun presided over the head, the moon over the right arm, Venus over the left, Jupiter over the stomach, Mars the parts below, Mercury over the right leg, and Saturn over the left.

Among the constellations, the Ram governed the head ; the Bull the neck ; the Twins the arms and shoulders ; the Crab the chest and the heart ; the Lion the stomach ; the abdomen corresponded to the sign of the Virgin ; the reins to the Balance ; then came the Scorpion ; the Archer, governing the thighs ; the He-goat the knees ; the Waterer the legs ; and the Fishes the feet.

Albert the Great assigned to the stars the following influences :—Saturn was thought to rule over life, changes, sciences, and buildings; Jupiter over honour, wishes, riches, and cleanness; Mars over war, prisons, marriages, and hatred; the sun over hope, happiness, gain, and heritages; Venus over friendships and amours; Mercury over illness, debts, commerce, and fear; the moon over wounds, dreams, and larcenies.

Each of these stars also presides over particular days of the week, particular colours, and particular metals.

The sun governed the Sunday; the moon, Monday; Mars, Tuesday; Mercury, Wednesday; Jupiter, Thursday; Venus, Friday; and Saturn, Saturday; which is partially indicated by our own names of the week, but more particularly in the French names, which are each and all derived from these stars.

The sun represented yellow; the moon, white; Venus, green; Mars, red; Jupiter, blue; Saturn, black; Mercury, shaded colours.

We have already indicated the metals that corresponded to each.

The sun was reckoned to be beneficent and favourable; Saturn to be sad, morose, and cold; Jupiter, temperate and benign; Mars, vehement; Venus, benevolent and fertile; Mercury, inconstant; and the moon, melancholy.

Among the constellations, the Ram, the Lion, and the Archer were hot, dry and vehement. The Bull, the Virgin, and the He-goat were heavy, cold, and dry; the Twins, the Balance,

PLATE XV.—AN ASTROLOGER AT WORK.

and the Waterer were light, hot, and moist; the Crab,
Scorpion, and the Fishes were moist, soft, and cold.

In this way the heavens were made to be intimately
connected with the affairs of earth ; and astrology was in
equally intimate connection with astronomy, of which it
may in some sense be considered the mother.   The drawers
of horoscopes were at one time as much in request as
lawyers or doctors.   One Thurneisen, a famous astrologer
and an extraordinary man, who lived last century at the
electoral court of Berlin, was at the same time physician,
chemist, drawer of horoscopes, almanack maker, printer, and
librarian.   His astrological reputation was so widespread
that scarcely a birth took place in families of any rank in
Germany, Poland, Hungary, or even England without there
being sent an immediate envoy to him to announce the
precise moment of birth.   He received often three and
sometimes as many as ten messages a day, and he was at
last so pressed with business that he was obliged to take
associates and agents.

In the days of Kepler we know that astrology was more
thought of than astronomy, for though on behalf of the
world he worked at the latter, for his own daily bread he
was in the employ of the former, making almanacks and
drawing horoscopes that he might live.

CHAPTER XIV.

TIME AND THE CALENDAR.

THE opinions of thinkers on the nature of time have been very varied. Some have considered time as an absolute reality, which is exactly measured by hours, days, and years, and is as known and real as any other object whose existence is known to us. Others maintain that time is only a matter of sensation, or that it is an illusion, or a hallucination of a lively brain.

The definitions given of it by different great writers is as various. Thus Kant calls it "one of the forms of sensibility." Schelling declares it is "pure activity with the negation of all being." Leibnitz defines it "the order of successions" as he defined space to be the order of co-existences. Newton and Clarke make space and time two attributes of the Deity.

A study of the astronomical phenomena of the universe, and a consideration of their teaching, give us authority for

saying, that neither space nor time are realities, but that
the only things absolute are eternity and infinity.

In fact, we give the name of time to the succession of
the terrestrial events measured by the motion of the earth.
If the earth were not to move, we should have no means
of measuring, and consequently no idea of time as we
have it now. So long as it was believed that the earth was
at rest, and that the sun and all the stars turned round us,
this apparent motion was then, as the real motion of the
earth is now, the method of generating time. In fact, the
Fathers said that at the end of the world the diurnal
motion would cease, and there would be no more time.
But let us examine the fact a little further.

Suppose for an instant that the earth was, as it was
formerly believed to be, an immense flat surface, which was
illuminated by a sun which remained always immovable at
the zenith, or by an invariable diffused light—such an
earth being supposed to be alone by itself in the universe
and immovable. Now if there were a man created on that
earth, would there be such a thing as "time" for him?
The light which illumines him is immovable. No moving
shadow, no gnomon, no sun-dial would be possible. No day
nor night, no morning nor evening, no year. Nothing that
could be divided into days, hours, minutes, and seconds.

In such a case one would have to fall back upon some
other terminating events, which would indicate a lapse of

time; such for instance as the life of a man. This, however, would be no universal measure, for on one planet the life might be a thousand years, and on another only a hundred.

Or we may look at it in another way. Suppose the earth were to turn twice as fast about itself and about the sun, the persons who lived sixty of such years would only have lived thirty of our present years, but they would have seen sixty revolutions of the earth, and, rigorously speaking, would have lived sixty years. If the earth turned ten times as fast, sixty years would be reduced to ten, but they would still be sixty of those years. We should live just as long; there would be four seasons, 365 days, &c., only everything would be more rapid: but it would be exactly the same thing for us, and the other apparently celestial motions having a similar diminution, there would be no change perceived by us.

Again, consider the minute animals that are observable under the microscope, which live but for five minutes. During that period, they have time to be born and to grow. From embryos they become adult, marry, so to speak, and have a numerous progeny, which they develop and send into the world. Afterwards they die, and all this in a few minutes. The impressions which, in spite of their minuteness, we are justified in presuming them to possess, though rapid and fleeting, may be as profound for them in proportion as ours are to us, and their measure of time would be very different from ours. All is relative. In absolute value, a life completed in a hundred years is not longer than one that is finished in five minutes.

It is the same for space. The earth has a diameter of 8,000 miles, and we are five or six feet high. Now if, by any process, the earth should diminish till it became as small as a marble, and if the different elements of the world underwent a corresponding diminution, our mountains might become as small as grains of sand, the ocean might be but a drop, and we ourselves might be smaller than the microscopic animals adverted to above. But for all that nothing would have changed for us. We should still be our five or six feet high, and the earth would remain exactly the same number of our miles.

A value then that can be decreased and diminished at pleasure without change is not a mathematical absolute value. In this sense then it may be said that neither time nor space have any real existence.

Or once again. Suppose that instead of our being on the globe, we were placed in pure space. What time should we find there? No time. We might remain ten years, twenty, a hundred, or a thousand years, but we should never arrive at the next year! In fact each planet makes its own time for its inhabitants, and where there is no planet or anything answering to it there is no time. Jupiter makes for its inhabitants a year which is equal to twelve years

of ours, and a day of ten of our hours. Saturn has a year
equal to thirty of ours, and days of ten hours and a quarter.
In other solar systems there are two or three suns, so that
it is difficult to imagine what sort of time they can have.
All this infinite diversity of time takes place in eternity,
the only thing that is real. The whole history of the earth
and its inhabitants takes place, not in time, but in eternity.
Before the existence of the earth and our solar system, there
was another time, measured by other motions, and having
relation to other beings. When the earth shall exist no
longer, there may be in the place we now occupy, another
time again, for other beings. But they are not realities. A
hundred millions of centuries, and a second, have the same
real length in eternity. In the middle of space, we could
not tell the difference. Our finite minds are not capable of
grasping the infinite, and it is well to know that our only
idea of time is relative, having relation to the regular events
that befall this planet in its course, and not a thing which
we can in any way compare with that which is so alarming
to the ideas of some—eternity.

We have then to deal with the particular form of time
that our planet makes for us, for our personal use.

It turns about the sun. An entire circuit forms a
period, which we can use for a measure in our terrestrial
affairs. We call it a year, or in Latin *annus*, signifying a
circle, whence our word *annual*.

A second, shorter revolution, turns the earth upon itself, and brings each meridian directly facing the sun, and then round again to the opposite side. This period we call a *day*, from the Latin *dies*, which in Italian becomes *giorne*, whence the French *jour*. In Sanscrit we have the same word in *dyaus*.

The length of time that it takes for the earth to arrive at the same position with respect to the stars, which is called a sidereal year, amounts to 365·2563744 days. But during this time, as we have seen, the equinox is displaced among the stars. This secular retrogression brings it each year a little to the east of its former position, so that the sun arrives there about eleven minutes too soon. By taking this amount from the sidereal we obtain the tropical year, which has reference to the seasons and the calendar. Its length is 365·2422166 days, or 365 days, 5 hours, 48 minutes, 47·8 seconds.

In what way was the primitive year regulated? was it a solar or a sidereal year?

There can be no doubt that when there was an absence of all civilisation and a calendar of any sort was unknown, the year meant simply the succession of seasons, and that no attempt would be made to reckon any day as its commencement. And as soon as this was attempted a difficulty would arise from there not being an exact number of days in the year. So that when reckoned as the interval between certain positions of the sun they would be of different lengths, which would introduce some difficulty as to the commencement of the year. Be this the case, however, or not, Mr. Haliburton's researches seem to show that the earliest form of year was the sidereal one, and that it was regulated by the Pleiades.

In speaking of that constellation we have noticed that among the islanders of the southern hemisphere and others there are two years in one of ours, the first being called the Pleiades above and the second the Pleiades below ; and we have seen how the same new year's day has been recognised in very many parts of the world and among the ancient Egyptians and Hindoos. This year would begin in November, and from the intimate relation of all the primitive calendars that have been discovered to a particular day, taken as November 17 by the Egyptians, it would appear probable that for a long time corrections were made both by the Egyptians and others in order to keep the phenomenon of the Pleiades just rising at sunset to one particular named day of their year —showing that the year they used was a sidereal one. This can be traced back as far as 1355 B.C. among the Egyptians, and to 1306 B.C. among the Hindoos. There seem to have been in use also shorter periods of three months, which, like the two-season year, appear to have been, as they are now among the Japanese, regulated by the different positions of the Pleiades.

Among the Siamese of the present day, there are both

forms of the year existing, one sidereal, beginning in November, and regulated by the fore-named constellation; and the other tropical, beginning in April. Whether, however, the year be reckoned by the stars or by the sun, there will always be a difficulty in arranging the length of the year, because in each case there will be about a quarter of a day over.

It seems, too, to have been found more convenient in early times to take 360 days as the length of the year, and to add an intercalary month now and then, rather than 365 and add a day.

Thus among the earliest Egyptians the year was of 360 days, which were reckoned in the months, and five days were added each year, between the commencement of one and the end of the other, and called unlucky days. It was the belief of the Egyptians that these five days were the birthdays of their principal gods; Osiris being born on the first, Anieris (or Apollo) on the second, Typhon on the third, Isis on the fourth, Nephys (or Aphrodite) on the fifth. These appear to have some relation with similar unlucky days among the Greeks and Romans, and other nations.

The 360 days of the Egyptian year were represented at Acantho, near Memphis, in a symbolical way, there being placed a perforated vessel, which each day was filled with water by one of a company of 360 priests, each priest having charge over one day in the year. A similar symbolism was used at the tomb of Osiris, around which were placed 360 pitchers, one of which each day was filled with milk.

On the other hand, the 365 days were represented by the tomb of Osymandyas, at Thebes, being surrounded by a circle of gold which was one cubit broad and 365 cubits in circumference. On the side were written the risings and settings of the stars, with the prognostications derived from them by the Egyptian astrologers. It was destroyed, however, by Cambyses when the Persians conquered Egypt.

They divided their year according to Herodotus into twelve months, the names of which have come down to us.

Even with the 365 days, which their method of reckoning would practically come to, they would still be a quarter of a day each year short; so that in four years it would amount to a whole day, an error which would amount to something perceptible even during the life of a single man, by its bringing the commencement of the civil year out of harmony with the seasons. In fact the first day of the year would gradually go through all the seasons, and at the end of 1460 solar years there would have been completed 1461 civil years, which would bring back the day to its original position. This period represents a cycle of years in which approximately the sun and the earth come to the same relative position again, as regards the earth's rotation on its axis and revolution round the sun. This cycle was noticed by Firmicius. Another more accurate cycle of the same kind, noticed by Syncellus, is obtained by multiplying 1461 by 25, making 36,525 years, which takes into account the defect which the extra hours over 365 have from six. The Egyptians, however, did not allow their year to get into so large an error, though it was in error by their using sidereal time, regulating their year, and intercalating days, first according to the risings of the Pleiades, and after according to that of Sirius, the dog-star, which announced to them the approaching overflowing of the Nile, a phenomenon of such great value to Egypt that they celebrated it with annual fêtes of the greatest magnificence.

Among the Babylonians, as we are informed by Mr. Sayce, the year was divided into twelve lunar months and 360 days, an intercalary month being added whenever a certain star, called the " star of stars," or Icu, also called Dilgan, by the ancient Accadians, meaning the " messenger of light," and what is now called Aldebaran, which was just in advance of the sun when it crossed the vernal equinox, was not parallel with the moon until the third of Nisan, that is, two days after the equinox. They also added shorter months of a few days each when this system became insufficient to keep their calendar correct.

They divided their year into four quarters of three months each ; the spring quarter not commencing with the beginning of the year when the sun entered the spring equinox, proving that the arrangement of seasons was subsequent to the settling of the calendar. The names of their months were given them from the corresponding signs of the zodiac ; which was the same as our own, though the zodiac began with Aries and the year with Nisan.

They too had cycles, but they arose from a very different cause ; not from errors of reckoning in the civil year or the revolution of the earth, but from the variations of the weather. Every twelve solar years they expected to have the same weather repeated. When we connect this with their observations on the varying brightness of the sun, especially at the commencement of the year on the first of Nisan, which they record at one time as " bright yellow " and at another as " spotted," and remember that modern researches have shown that weather is certainly in some way dependent on the solar spots, which have a period *now* of about eleven years, we cannot help fancying that they were very near to making these discoveries.

The year of the ancient Persians consisted of 365 days. The extra quarter of a day was not noticed for 120 years, at the end of which they intercalated a month—in the first instance, at the end of the first month, which was thus doubled. At the end of another 120 years they inserted an intercalary month after the second month, and so on through all their twelve months. So that after 1440 years the series began again. This period they called the intercalary cycle.

The calendar among the Greeks was more involved, but

more useful.   It was *luni-solar*, that is to say, they regulated it at the same time by the revolutions of the moon and the motion about the sun, in the following manner :—

The year commenced with the new moon nearest to the 20th or 21st of June, the time of the summer solstice; it was composed in general of twelve months, each of which commenced on the day of the new moon, and which had alternately twenty-nine and thirty days.

This arrangement, conformable to the lunar year, only gave 354 days to the civil year, and as this was too short by ten days, twenty-one hours, this difference, by accumulation, produced nearly eighty-seven days at the end of eight years, or three months of twenty-nine days each.   To bring the lunar years into accordance with the solstices, it was necessary to add three intercalary months every eight years.

The phases of the moon being thus brought into comparison with the rotation of the earth, a cycle was discovered by Meton, now known as the Metonic cycle, useful also in predicting eclipses, which comprised nineteen years, during which time 235 lunations will have very nearly occurred, and the full moons will return to the same dates.   In fact, the year and the lunation are to one another very nearly in the proportion of 235 to 19.   By observing for nineteen years the positions and phases of the moon, they will be found to return again in the same order at the same times, and they can therefore be predicted.   This lunar cycle was adopted in the year 433 B.C. to regulate the luni-solar calendar, and it was engraved in letters of gold on the walls of the temple of Minerva, from whence comes the name *golden number*, which is given to the number that marks the place of the given year in this period of nineteen.

Caliphus made a larger and more exact cycle by multiplying by four and taking away one day. Thus he made of 27,759 days 76 Julian years, during which there were 940 lunations.

The Roman calendar was even more complicated than the Greek, and not so good. Romulus is said to have given to his subjects a strange arrangement that we can no longer understand. More of a warrior than a philosopher, this founder of Rome made the year to consist of ten months, some being of twenty days and others of fifty-five. These unequal lengths were probably regulated by the agricultural works to be done, and by the prevailing religious ideas. After the conclusion of these days they began counting again in the same order; so that the year had only 304 days.

The first of these ten months was called *Mars* after the name of the god from whom Romulus pretended to have descended. The name of the second, Aprilis, was derived from the word *aperire*, to open, because it was at the time that the earth opened; or it may be, from Aphrodite, one of the names of Venus, the supposed grandmother of Æneas. The third month was consecrated to *Maïa*, the mother of

Mercury.   The names of the six others expressed simply
their order—Quintilis, the fifth ; Sextilis, the sixth ; September, the seventh ; and so on.

Numa added two months to the ten of Romulus ; one took
the name of *Januarius*, from *Janus :* the name of the other
was derived either from the sacrifices (*februalia*), by which
the faults committed during the course of the past year were
expiated, or from *Februo*, the god of the dead, to which the
last month was consecrated.   The year then had 355 days.

These Roman months have become our own, and hence a
special interest attaches to the consideration of their origin,
and the explanation of the manner in which they have been
modified and supplemented.   Each of them was divided into
unequal parts, by the days which were known as the calends,
nones, and ides.   The calends were invariably fixed to the
first day of each month ; the nones came on the 5th or 7th,
and the ides the 13th or 15th.

The Romans, looking forward, as children do to festive
days, to the fête which came on these particular days, named
each day by its distance from the next that was following.
Immediately after the calends of a month, the dates were
referred to the nones, each day being called seven, six, five,
and so on days before the nones; on the morrow of the
nones they counted to the ides; and so the days at the
end of the month always bore the name of the calends of
the month following.

To complete the confusion the 2nd day before the fête was called the 3rd, by counting the fête itself as the 1st, and so they added one throughout to the number that *we* should now say expressed our distance from a certain date.

Since there were thus ten days short in each year, it was soon found necessary to add them on, so a supplementary month was created, which was called Mercedonius. This month, by another anomaly, was placed between the 23rd and 24th of February. Thus, after February 23rd, came 1st, 2nd, 3rd of Mercedonius; and then after the dates of this supplementary month were gone through, the original month was taken up again, and they went on with the 24th of February.

And finally, to complete the medley, the priests who had the charge of regulating this complex calendar, acquitted themselves as badly as they could; by negligence or an arbitrary use of their power they lengthened or shortened the year without any uniform rule. Often, indeed, they consulted in this nothing but their own convenience, or the interests of their friends.

The disorder which this license had introduced into the calendar proceeded so far that the months had changed from the seasons, those of winter being advanced to the autumn, those of the autumn to the summer. The fêtes were celebrated in seasons different from those in which they were instituted, so that of Ceres happened when the wheat was in the blade, and that of Bacchus when the raisins were green. Julius Cæsar, therefore, determined to establish a solar year according to the known period of revolution of the sun, that is 365 days and a quarter. He ordained that each fourth year a day should be intercalated in the place where the month Mercedonius used to be inserted, *i.e.* between the 23rd and 24th of February.

The 6th of the calends of March in ordinary years was the 24th of February; it was called *sexto-kalendas.* When an extra day was put in every fourth year before the 24th, this was a second 6th day, and was therefore called *bissexto-kalendas,* whence we get the name bissextile, applied to leap year.

But it was necessary also to bring back the public fêtes to the seasons they ought to be held in : for this purpose Cæsar was obliged to insert in the current year, 46 B.C. (or 708 A.U.C.), two intercalary months beside the month Mercedonius. There was, therefore, a year of fifteen months divided into 445 days, and this was called the year of confusion.

Cæsar gave the strictest injunctions to Sosigenes, a celebrated Alexandrian astronomer whom he brought to Rome for this purpose; and on the same principles Flavius was ordered to compose a new calendar, in which all the Roman fêtes were entered—following, however, the old method of reckoning the days from the calends, nones, and ides. Antonius, after the death of Cæsar, changed the name of Quintilis, in which Julius Cæsar was born, into the name

Julius, whence we derive our name July.   The name of
Augustus was given to the month *Sextilis,* because the Emperor
Augustus obtained his greatest victories during that month.

Tiberius, Nero, and other imperial monsters attempted
to give their names to the other months.   But the people
had too much independence and sense of justice to accord
them such a flattery.

The remaining months we have as they were named in
the days of Numa Pompilius.

FIG. 57.—THE ROMAN CALENDAR.

A cubical block of white marble has been found at
Pompeii which illustrates this very well.

Each of the four sides is divided into three columns, and
on each column is the information about the month.   Each
month is surmounted by the sign of the zodiac through
which the sun is passing.   Beneath the name of the month
is inscribed the   number   of days it contains; the date
of the nones, the number of the hours of the day, and of

the night; the place of the sun, the divinity under whose protection the month is placed, the agricultural works that are to be done in it, the civil and ecclesiastical ceremonies that are to performed. These inscriptions are to be seen under the month January to the left of the woodcut.

The reform thus introduced by Julius Cæsar is commonly known as the *Julian reform*. The first year in which this calendar was followed was 44 B.C.

The Julian calendar was in use, without any modification, for a great number of years; nevertheless, the mean value which had been assigned to the civil year being a little different to that of the tropical, a noticeable change at length resulted in the dates in which, each year, the seasons commenced; so that if no remedy had been introduced, the same season would be displaced little by little each year, so as to commence successively in different months.

The Council of Nice, which was held in the year 325 of the Christian era, adopted a fixed rule to determine the time at which Easter falls. This rule was based on the supposed fact that the spring equinox happened every year on the 21st of March, as it did at the time of the meeting of the Council. This would indeed be the case if the mean value of the civil year of the Julian calendar was exactly equal to the tropical year. But while the first is 365·25 days, the second is 365·242264 days; so that the tropical year is too small by 11 minutes and 8 seconds. It follows hence that after the lapse of four Julian years the vernal equinox, instead of happening exactly at the same time as it did four years before, will happen 44 minutes 32 seconds too soon; and will gain as much in each succeeding four years. So that at the end of a certain number of years, after the year 325, the equinox will happen on the 20th of March, afterwards on the 19th, and so on. This continual advance notified by the astronomers, determined Pope Gregory XIII. to introduce a new reform into the calendar.

It was in the year 1582 that the *Gregorian reform* was put into operation. At that epoch the vernal equinox happened on the 11th instead of the 21st of March. To get rid of this advance of ten days that the equinox had made and to bring it back to the original date, Pope Gregory decided that the day after the 4th of October, 1582, should be called the 15th instead of the 5th. This change only did away with the inconvenience at the time attaching to the Julian calendar; it was necessary to make also some modification in the rule which served to determine the lengths of the civil years, in order to avoid the same error for the future.

So the Pope determined that in each 400 years there should be only 97 bissextile years, instead of 100, as there used to be in the Julian calendar. This made three days taken off the 400 years, and in consequence the mean value of the civil year is reduced to 365·2425 days, which is not far from the true tropical year. The Gregorian year thus obtained is still too great by ·000226 of a day ; the date of the vernal equinox will still then advance in virtue of this excess, but it is easy to see that the Gregorian reform will suffice for a great number of centuries.

The method in which it is carried out is as follows :—In the Julian calendar each year that divided by four when expressed in its usual way, by A.D., was a leap year, and therefore each year that completed a century was such, as 1300, 1400 and so on—but in the Gregorian reform, all these century numbers are to be reckoned common years, unless the number without the two cyphers divides by four ; thus 1,900 will be a common year and 2,000 a leap year. It is easy to see that this will leave out three leap years in every 400 years.

The Gregorian calendar was immediately adopted in France and Germany, and a little later in England. Now it is in operation in all the Christian countries of Europe, except Russia, where the Julian calendar is still followed. It follows that Russian dates do not agree with ours. In 1582, the difference was ten days, and this difference remained the same till the end of the seventeenth century, when the year 1700 was bissextile in the Julian, but not in the Gregorian calendar, so the difference increased to eleven days, and now in the same way is twelve days.

Next to the year, comes the day as the most natural division of time in connection with the earth, though it admits of less difference in its arrangements, as we cannot be mistaken as to its length. It is the natural standard too of our division of time into shorter intervals such as hours, minutes, and seconds. By the word *day* we mean of course the interval during which the earth makes a complete revolution round itself, while *daytime* may be used to express the portion of it during which our portion of the earth is towards the sun. The Greeks to avoid ambiguity used the word *nyctemere*, meaning night and day.

No ancient nation is known that did not divide the day into twenty-four hours, when they divided it at all into such small parts, which seems to show that such a division was comparatively a late institution, and was derived from the invention of a single nation. It would necessarily depend on the possibility of reckoning shorter periods of time than the natural one of the day. In the earliest ages, and even afterwards, the position of the sun in the heavens by day, and the position of the constellations by night, gave approximately the time. Instead of asking What " o'clock " is it ? the Greeks would say, " What star is passing ? " The next method of determining time depended on the uniform motion of water from a cistern. It was invented by the Egyptians, and was called a clepsydra, and was in use among the Babylonians, the Greeks, and the Romans. The more accurate measurement of time by means of clocks was not introduced till about 140 B.C., when Trimalcion had one in his dining chamber. The use of them, however, had been so lost that in 760 A.D. they were considered quite novelties. The clocks, of course, have to be regulated by the sun, an operation which has been the employment of astronomers, among other things, for centuries. Each locality had its own time according to the moment when the sun passed the meridian of the place, a moment which was determined by observation.

Before the introduction of the hour, the day and night appear to have been divided into watches. Among the Babylonians the night was reckoned from what we call 6 A.M. to 6 P.M., and divided into three watches of four hours each—called the "evening," "middle," and "morning" watch. These were later superseded by the more accurate hour, or rather "double hour" or *casbri*, each of which was divided into sixty minutes and sixty seconds, and the change taking place not earlier than 2,000 B.C. Whether the Babylonians (or Accadians) were the inventors of the hour it is difficult to say, though they almost certainly were of other divisions of time. It is remarkable that in the ancient Jewish Scriptures we find no mention of any such division until the date at which the prophecy of Daniel was written, that is, until the Jews had come in contact with the Babylonians.

Some nations have counted the twenty-four hours consecutively from one to twenty-four as astronomers do now,

but others and the majority have divided the whole period
into two of twelve hours each.

The time of the commencement of the day has varied
much with the different nations.

The Jews, the ancient Athenians, the Chinese, and
several other peoples, more or less of the past, have commenced the day with the setting of the sun, a custom which
perhaps originated with the determination of the commencement of the year, and therefore of the day, by the observation
of some stars that were seen at sunset, a custom continued
in our memory by the well-known words, "the evening and
the morning were the first day."

The Italians, till recently, counted the hours in a single
series, between two settings of the sun. The only gain in
such a method would be to sailors, that they might know
how many hours they had before night overtook them; the
sun always setting at twenty-four o'clock; if the watch
marked nineteen or twenty, it would mean they had five or
four hours to see by—but such a gain would be very small
against the necessity of setting their watches differently
every morning, and the inconvenience of never having
fixed hours for meals.

Among the Babylonians, Syrians, Persians, the modern
Greeks, and inhabitants of the Balearic Isles, &c., the day
commenced with the rising of the sun. Nevertheless, among
all the astronomical phenomena that may be submitted to

observation, none is so liable to uncertainty as the rising and setting of the heavenly bodies, owing among other things to the effects of refraction.

Among the ancient Arabians, followed in this by the author of the *Almagesta*, and by Ptolemy, the day commenced at noon. Modern astronomers adopt this usage. The moment of changing the date is then always marked by a phenomenon easy to observe.

Lastly, that we may see how every variety possible is sure to be chosen when anything is left to the free choice of men, we know that with the Egyptians, Hipparchus, the ancient Romans, and all the European nations at present, the day begins at midnight. Copernicus among the astronomers of our era followed this usage. We may remark that the commencement of the astronomical day commences twelve hours *after* the civil day.

Of the various periods composed of several days, the week of seven days is the most widely spread—and of considerable antiquity. Yet it is not the universal method of dividing months. Among the Egyptians the month was divided into periods of ten days each; and we find no sign of the seven days—the several days of the whole month having a god assigned to each. Among the Hindoos no trace has been found by Max Müller in their ancient Vedic literature of any such division, but the month is divided into two according to the moon; the *clear* half from the new to the full moon, the *obscure* half from the full to the new, and a similar division has been found among the Aztecs. The Chinese divide the month like the Egyptians. Among the Babylonians two methods of dividing the month existed, and both of them from the earliest times. The first method was to separate it into two halves of fifteen days each, and each of these periods into three shorter ones of five days, making six per month. The other method is the week of seven days. The days of the week with them, as they are with many nations now, were named after the sun and moon and the five planets, and the 7th, 14th, 19th, 21st, and 28th days of each month—days separated by seven days each omitting the 19th—were termed " days of rest," on which certain works were forbidden to be done. From this it is plain that we have here all the elements of our modern week. We find it, as is well known, in the earliest of Hebrew writings, but without the mark which gives reason for the number seven, that is the names of the seven heavenly bodies. It would seem most probable, then, that we must look to the Accadians as the originators of our modern week, from whom the Hebrews may have—and, if so, at a very early period—borrowed the idea.

It is known that the week was not employed in the ancient calendars of the Romans, into which it was afterwards introduced through the medium of the biblical traditions, and became a legal usage under the first Christian Emperors.

From thence it has been propagated together with the Julian calendar amongst all the populations that have been subjected to the Roman power. We find the period of seven days employed in the astronomical treatises of Hindoo writers, but not before the fifth century.

Dion Cassius, in the third century, represents the week as universally spread in his times, and considers it a recent invention which he attributes to the Egyptians; meaning thereby, doubtless, the astrologers of the Alexandrian school, at that time very eager to spread the abstract speculations of Plato and Pythagoras.

If the names of the days of the week were derived from the planets, the sun and moon, as is easy to see, it is not so clear how they came to have their present order. The original order in which they were supposed to be placed in the various heavens that supported them according to their distance from the earth was thus :—Saturn, Jupiter, Mars, the Sun, Venus, Mercury, the Moon. One supposition is that each hour of the day was sacred to one of these, and that each day was named from the god that presided over the first hours. Now, as seven goes three times into twenty-four, and leaves three over, it is plain that if Saturn began the first hour of Saturday, the next day would begin with the planet three further on in the series, which would bring us to the Sun for Sunday, three more would bring us next day to the Moon for Monday, and so to Mars for Tuesday, to Mercury

for Wednesday, to Jupiter for Thursday, to Venus for Friday,
and so round again to Saturn for Saturday.

The same method is illustrated by putting the symbols in
order round the circumference of a circle, and joining them
by lines to the one most opposite, following always in the
same order as in the following figure.   We arrive in this way
at the order of the days of the week.

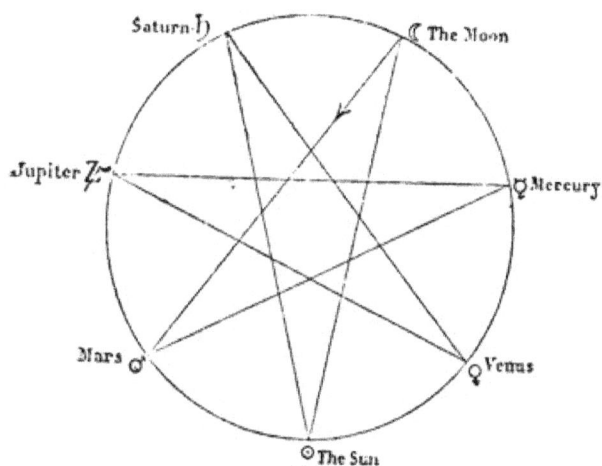

FIG. 58.

All the nations who have adopted the week have not kept
to the same names for them, but have varied them according
to taste. Thus Sunday was changed by the Christian Church
to the " Lord's Day," a name it still partially retains among
ourselves, but which is the regular name among several

continental nations, including the corrupted *Dimanche* of the French. The four middle days have also been very largely changed, as they have been among ourselves and most northern nations to commemorate the names of the great Scandinavian gods Tuesco, Woden, Thor, and Friga. This change was no doubt due to the old mythology of the Druids being amalgamated with the new method of collecting the days into weeks.

We give below a general table of the names of the days of the week in several different languages.

English.	French.	Italian.	Spanish.	Portuguese.
Sunday.	Dimanche.	Domenica.	Domingo.	Domingo.
Monday.	Lundi.	Lunedi.	Luneo.	Secunda feira.
Tuesday.	Mardi.	Marteti.	Martes.	Terça feira.
Wednesday.	Mercredi.	Mercoledi.	Miercoles.	Quarta feira.
Thursday.	Jeudi.	Giovedi.	Jueves.	Quinta feira.
Friday.	Vendredi.	Venerdi.	Viernes.	Sexta feira.
Saturday.	Samedi.	Sabbato.	Sabado.	Sabbado.

German.	Anglo-Saxon.	Ancient Frisian.	Ancient Northmen.	Dutch.
Sonntag.	Sonnan dag.	Sonna dei.	Sunnu dagr.	Zondag.
Montag.	Monan dag.	Mona dei.	Mána dagr.	Maandag.
Dienstag.	Tives dag.	Tys dei.	Tvrs dagr.	Dingsdag.
Mitwoch.	Vodenes dag.	Werns dei.	Odins dagr.	Woensdag.
Donnerstag.	Thunores dag.	Thunres dei.	Thors dagr.	Donderdag.
Freitag.	Frige dag.	Frigen dei.	Fria dagr.	Vrijdag.
Samstag.	Soeternes dag.	Sater dei.	{Laugar dagr } {(washing day)}	Zaturdag.

The cycle which must be completed with the present calendar to bring the same day of the year to the same day of the week, is twenty-eight years, since there is one day over every ordinary year, and two every leap year; which will make an overlapping of days which, except at the centuries, will go through all the changes in twenty-eight times, which forms what is called the solar cycle.

There is but one more point that will be interesting about the calendar, namely, the date from which we reckon our years.

Among the Jews it was from the creation of the world, as recorded in their sacred books—but no one can determine when that was with sufficient accuracy to make it represent anything but an agreement of the present day. Different interpreters do not come within a thousand years of one another for its supposed date; although some of them have determined it very accurately to their own satisfaction—one going so far as to say that creation finished at nine o'clock one Sunday morning! In other cases the date has been reckoned from national events—as in the Olympiads, the foundation of Rome, &c. The word we now use, ÆRA, points to a particular date from which to reckon, since it is composed of the initials of the words AB EXORDIO REGNI AUGUSTI "from the commencement of the reign of Augustus." At the present day the point of departure, both forwards and backwards, is the year of the birth of Jesus Christ—a date which is itself controverted, and the use of which did not exist among the first Christians. They exhibited great indifference, for many centuries, as to the year in which Jesus Christ entered the world. It was a monk who lived in obscurity at Rome, about the year 580, who was a native of so unknown a country that he has been called a Scythian, and whose name was Denys, surnamed *Exiguus*, or the Little, who first attempted to discover by chronological calculations the year of the birth of Jesus Christ.

The era of Denys the Little was not adopted by his contemporaries. Two centuries afterwards, the Venerable Bede exhorted Christians to make use of it—and it only came into general use about the year 800.

Among those who adopted the Christian era, some made the year commence with March, which was the first month of the year of Romulus; others in January, which commences the year of Numa; others commenced on Christmas Day; and others on Lady Day, March 25. Another form of nominal year was that which commenced with Easter Day, in which case, the festival being a movable one, some years were shorter than others, and in some years there might be two 2nd, 3rd, &c., of April, if Easter fell in one year on the 2nd, and next year a few days later.

The 1st of January was made to begin the year in Germany in 1500. An edict of Charles IX. prescribes the same in France in 1563. But it was not till 1752 that the change was made in England by Lord Chesterfield's Act. The year 1751, as the year that had preceded it, began on

March 25th, and it should have lasted till the next Lady
Day; but according to the Act, the months of January,
February, and part of March were to be reckoned as part
of the year 1752. By this means the unthinking seemed
to have grown old suddenly by three months, and popular
clamour was raised against the promoter of the Bill, and
cries raised of "Give us our three months." Such have
been the various changes that our calendar has undergone
to bring it to its present state.

CHAPTER XV.

PERHAPS the most anxious question that has been asked of the astronomer is when the world is to come to an end. It is a question which, of course, he has no power to answer with truth ; but it is also one that has often been answered in good faith. It has perhaps been somewhat natural to ask such a question of an astronomer, partly because his science naturally deals with the structure of the universe, which might give some light as to its future, and partly because of his connection with astrology, whose province it was supposed to be to open the destiny of all things. Yet the question has been answered by others than by astronomers, on grounds connected with their faith. In the early ages of the Church, the belief in the rapid approach of the end of the world was universally spread amongst Christians. The Apocalypse of St. John and the Acts of the Apostles seemed to announce its coming before that generation passed away. Afterwards, it was expected at the year 1000 ; and though these beliefs did not rest in any way on astronomical grounds, yet to that science was recourse had for encouragement or discouragement of the idea. The middle ages, full of simple faith and superstitious credulity, were filled with fear of this terrible catastrophe.

As the year 1000 approached, the warnings became frequent and very pressing. Thus, for example, Bernard of Thuringia, about 960, began to announce publicly that the world was about to end, declaring that he had had a particular revelation of the fact. He took for his text the enigmatical words of the Apocalypse: "At the end of one thousand years, Satan shall be loosed from his prison, and shall seduce the people that are in the four quarters of the earth. The book of life shall be open, and the sea shall give up her dead." He fixed the day when the Annunciation of the Virgin should coincide with Good Friday as the end of all things. This happened in 992, but nothing extraordinary happened.

During the tenth century the royal proclamations opened by this characteristic phrase: *Whereas the end of the world is approaching.* . . . .

In 1186 the astrologers frightened Europe by announcing a conjunction of all the planets. Rigord, a writer of that period, says in his *Life of Philip Augustus :* "The astrologers of the East, Jews, Saracens, and even Christians, sent letters all over the world, in which they predicted, with perfect assurance, that in the month of September there would be great tempests, earthquakes, mortality among men, seditions and discords, revolutions in kingdoms, and the destruction of all things. But," he adds, "the event very soon belied their predictions."

Some years after, in 1198, another alarm of the end of the world was raised, but this time it was not dependent on celestial phenomena. It was said that Antichrist was born in Babylon, and therefore all the human race would be destroyed.

It would be a curious list to make of all the years in which it was said that Antichrist was born; they might be counted by hundreds, to say nothing of the future.

At the commencement of the fourteenth century, the alchemist Arnault of Villeneuve announced the end of the world for 1335. In his treatise *De Sigillis* he applies the influence of the stars to alchemy, and expounds the mystical formula by which demons are to be conjured.

St. Vincent Ferrier, a famous Spanish preacher, gave to the world as many years' duration as there were verses in the Psalms—about 2537.

The sixteenth century produced a very plentiful crop of predictions of the final catastrophe. Simon Goulart, for example, gave the world an appalling account of terrible sights seen in Assyria—where a mountain opened and showed a scroll with letters of Greek—"The end of the world is coming." This was in 1532; but after that year had passed in safety, Leovitius, a famous astrologer, predicted it again

for 1584. Louis Gayon reports that the fright at this time
was great. The churches could not hold those who sought
a refuge in them, and a great number made their wills,
without reflecting that there was no use in it if the whole
world was to finish.

One of the most famous mathematicians of Europe, named
Stofller, who flourished in the 16th century, and who worked
for a long time at the reform of the calendar proposed by
the Council of Constance, predicted a universal deluge for
1524. This deluge was to happen in the month of February,
because Saturn, Jupiter, and Mars were then together in
the sign of the Fishes. Everyone in Europe, Asia, and Africa,
to whom these tidings came, was in a state of consternation.
They expected a deluge, in spite of the rainbow. Many
contemporary authors report that the inhabitants of the
maritime provinces of Germany sold their lands for a mere
trifle to those who had more money and less credulity.
Each built himself a boat like an ark. A doctor of
Toulouse, named Auriol, made a very large ark for himself,
his family, and his friends, and the same precautions were
taken by a great many people in Italy. At last the month
of February came, and not a drop of rain fell. Never
was a drier month or a more puzzled set of astrologers.
Nevertheless they were not discouraged nor neglected for
all that, and Stoffler himself, associated with the celebrated
Regiomontanus, predicted once more that the end of the

world would come in 1588, or at least that there would
be frightful events which would overturn the earth.

This new prediction was a new deception; nothing
extraordinary occurred in 1588. The year 1572, however,
witnessed a strange phenomenon, capable of justifying all
their fears. An unknown star came suddenly into view in
the constellation of Cassiopeia, so brilliant that it was
visible even in full daylight, and the astrologers calculated
that it was the star of the Magi which had returned, and
that it announced the second coming of Jesus Christ.

The seventeenth and eighteenth centuries were filled with
new predictions of great variety.

Even our own century has not been without such. A
religious work, published in 1826, by the Count Sallmard
Montfort, demonstrated perfectly that the world had no
more than ten years to exist. "The world," he said, "is
old, and its time of ending is near, and I believe that the
epoch of that terrible event is not far off. Jacob, the
chief of the twelve tribes of Israel, and consequently of the
ancient Church, was born in 2168 of the world, i.e., 1836
B.C. The ancient Church, which was the figure of the
new, lasted 1836 years. Hence the new one will only last
till 1836 A.D."

Similar prophecies by persons of various nations have
in like manner been made, without being fulfilled. Indeed,
we have had our own prophets, but they have proved

themselves incredulous of their own predictions, by taking leases that should *commence* in the year of the world's destruction.

But we have one in store for us yet. In 1840, Pierre Louis of Paris calculated that the end would be in 1900, and he calculated in this way:—The Apocalypse says the Gentiles shall occupy the holy city for forty-two months. The holy city was taken by Omar in 636. Forty-two months of years is 1260, which brings the return of the Jews to 1896, which will precede by a few years the final catastrophe. Daniel also announces the arrival of Antichrist 2,300 days after the establishment of Artaxerxes on the throne of Persia, 400 B.C., which again brings us to 1900.

Some again have put it at 2000 A.D., which will make 6,000 years, as they think, from the creation; these are the days of work; then comes the 1,000 years of millennial sabbath.

We are led far away by these vain speculations from the wholesome study of astronomy; they are useful only in showing how by a little latitude that science may wind itself into all the questions that in any way affect the earth.

Indeed, since the world began, the world will doubtless end, and astronomers are still asked how could it be brought about?

Certainly it is not an impossible event, and there are only too many ways in which it has been imagined it might occur.

The question is one that stands on a very different footing from that it occupied before the days of Galileo and Copernicus. *Then* the earth was believed to be the centre of the universe, and all the heavens and stars created for it. *Then* the commencement of the world was the commencement of the universe, its destruction would be the destruction of all. *Now*, thanks to the revolution in feeling that has been accomplished by the progress of astronomy, we have learned our own insignificance, and that amongst the infinite number of stars, each supporting their own system of inhabited planets, our earth occupies an infinitesimally small portion, and the destruction of it would make no difference whatever—still less its becoming uninhabitable. It is an event which must have happened and be happening to other worlds, without affecting the infinite life of the universe in any marked degree.

Nevertheless, for ourselves, the question remains as interesting as if we were the all in all, but must be approached in a different manner.

Numerous hypotheses have been put forth on the question but they may mostly be dismissed as vain.

Buffon calculated that it had taken 74,832 years for the earth to cool down to its present temperature, and that it will take 93,291 years more before it would be too cold for men to live upon it.   But Sir William Thomson has shown that the internal heat of the earth, supposed to be due to its cooling from fusion, cannot have seriously modified climate for a long series of years, and that life depends essentially on the heat of the sun.

Another hypothesis, the most ancient of all, is that which supposes the earth will be destroyed by fire.   It comes down from Zoroaster and the Jews ; and on the improbable supposition of the thin crust of the earth over a molten mass, this is thought possible.   However, as the tendency in the past has been all the other way, namely, to make the effect of the inner heat of the earth less marked on the surface, we have no reason to expect a reversal.

A third theory would make the earth die more gradually and more surely.   It is known that by the wearing down of the surface by the rains and rivers, there is a tendency to reduce mountains and all high parts of the earth to a uniform level, a tendency which is only counteracted by some elevating force within the earth.   If these elevating forces be supposed to be due to the internal heat—a hypothesis which cannot be proved—then with the cooling of the earth the elevating forces would cease, and, finally, the whole of the continent would be brought beneath the sea and terrestrial life perish.

Another interesting but groundless hypothesis is that of

Adhémar on the periodicity of deluges. This theory depends on the fact of the unequal length of the seasons in the two hemispheres. Our autumn and our winter last 179 days. In the southern hemisphere they last 186 days. These seven days, or 168 hours, of difference, increase each year the coldness of the pole. During 10,500 years the ice accumulates at one pole and melts at the other, thereby displacing the earth's centre of gravity. Now a time will arrive when, after the maximum of elevation of temperature on one side, a catastrophe will happen, which will bring back the centre of gravity to the centre of figure, and cause an immense deluge. The deluge of the north pole was 4,200 years ago, therefore the next will be 6,300 hence. It is very obvious to ask on this—*Why* should there be a *catastrophe?* and why should not the centre of gravity return *gradually* as it was gradually displaced?

Another theory has been that it would perish by a comet. That it will not be by the shock we have already seen from the light weight of the comet and from experience; but it has been suggested that the gas may combine with the air, and an explosion take place that would destroy us all; but is not that also contradicted by experience?

Another idea is that we shall finally fall into the sun by the resistance of the ether to our motion. Encke's comet loses in thirty-three years a thousandth part of its velocity. It appears then that we should have to wait millions of centuries before we came too near the sun.

In reality, however, we are simply dependent on our sun, and our destiny depends upon that.

In the first place, in its voyage through space it might encounter or come within the range of some dark body we at present know nothing of, and the attraction might put out of harmony all our solar system with calamitous results. Or since we are aware that the sun is a radiating body giving out its heat on all sides, and therefore growing colder, it may one day happen that it will be too cold to sustain life on the earth. It is, we know, a variable star, and stars have been seen to disappear, or even to have a catastrophe happen to them, as the kindling of enormous quantities of gas. A catastrophe in the sun will be our own end.

Fontenelle has amusingly described in verse the result of the sun growing cold, which may be thus Englished :—

> " Of this, though, I haven't a doubt,
> 　One day when there isn't much light,
> The poor little sun will go out
> 　And bid us politely—good-night.
> Look out from the stars up on high,
> 　Some other to help you to see ;
> I can't shine any longer, not I,
> 　Since shining don't benefit me.

> " Then down on our poor habitation
>   What numberless evils will fall,
>   When the heavens demand liquidation,
>   Why all will go smash, and then all
>   Society come to an end.
>   Soon out of the sleepy affair
>   His way will each traveller wend,
>   No testament leaving, nor heir."

The cooling of the sun must, however, take place very gradually, as no cooling has been perceived during the existence of man ; and the growth of plants in the earliest geological ages, and the life of animals, prove that for so long a time it has been within the limits within which life has been possible—and we may look forward to as long in the future.

It is not of course the time when the sun will become a dark ball, surrounded by illuminated planets, that we must reckon as the end of the earth. Life would have ceased long before that stage—no man will witness the death of the sun.

The diminution of the sun's heat would have for its natural effect the enlargement of the glacial zones! the sea and the land in those parts of the earth would cease to support life, which would gradually be drawn closer to the equatorial belt. Man, who by his nature and his intelligence is best fitted to withstand cold climates, would remain among the last of the inhabitants, reduced to the most

PLATE XVI.—THE END OF THE WORLD.

miserable nourishment. Drawn together round the equator, the last of the sons of earth would wage a last combat with death, and exactly as the shades approached, would the human genius, fortified by all the acquirements of ages past — give out its brightest light, and attempt in vain to throw off the fatal cover that was destined to engulf him. At last, the earth, fading, dry, and sterile, would become an immense cemetery. And it would be the same with the other planets. The sun, already become red, would at last become black, and the planetary system would be an assemblage of black balls revolving round a larger black ball.

Of course this is all imaginary, and cannot affect ourselves, but the very idea of it is melancholy, and enough to justify the words of Campbell :—

> " For this hath science searched on weary wing
> By shore and sea—each mute and living thing,
> Or round the cope her living chariot driven
> And wheeled in triumph through the signs of heaven.
> Oh, star-eyed science, hast thou wandered there
> To waft us home the message of despair ?"

In reality, as we know nothing of the origin, so we know nothing of the end of the world; and where so much has been accomplished, there are obviously infinite possibilities enough to satisfy the hopes of every one.

While some stars may be fading, others may be rising into their place, and man need not be identified with one earth alone, but may rest content in the idea that the life universal is eternal.

THE END.

LONDON: F. CLAY, SONS, AND TAYLOR, PRINTERS.

www.ingramcontent.com/pod-product-compliance
Lightning Source LLC
Chambersburg PA
CBHW021342210326
41599CB00011B/720